A history of the discoveries about the sexuality of the honey bee

Frédéric Eggers de Villepin

NB

Northern Bee Books

Published in the United Kingdom by
Northern Bee Books,
Scout Bottom Farm,
Mytholmroyd,
West Yorkshire HX7 5JS
Tel: 01422 882751
Email: jerry@northernbeebooks.co.uk

www.northernbeebooks.co.uk

ISBN 978-1-914934-88-9

Design and artwork, DM Design and Print

Cover: Abeille mère & organes intérieurs (*Mother bee & inner organs*)
- L'Apiculteur, XXXII (1889), 1, 10.

A history of the discoveries about the sexuality of the honey bee

by Frédéric Eggers de Villepin

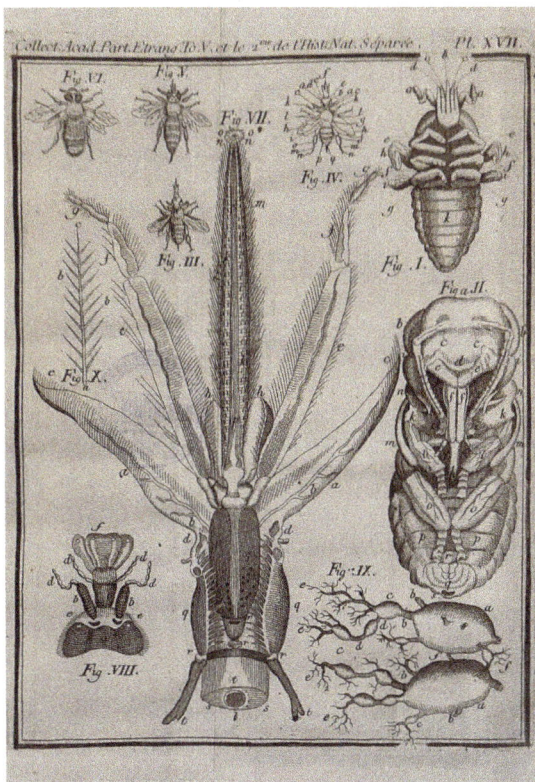

Jan Swammerdam, "Histoire des insectes", Plate XVII
- Honey bee details (SWAMMERDAM, 1758)

"It must be admitted, however, that there were considerable difficulties to be overcome in order to learn about the way in which bees are generated, but we had to confine ourselves to saying that we did not know until we had observations suitable for instruction. ... Nevertheless, there was still a sure way of determining at least the sex of each type of bee; & a way to which Swammerdam did not fail to have recourse, dissection; of examining the inner parts of the different types of flies in a hive."

Réaumur, 'Mémoire pour servir à l'histoire des insectes' (A memoir of the history of insects), 1740.

Contents

Introduction

Fig. 1 *Apiarium, the first publication of drawings based on microscopic views.*
Federico Cesi and Francesco Stelluti, 1625.

Putting together this history of the discoveries about the sexuality of the honey bee gives us the opportunity to delve into a past where intuitions, advances, sometimes lasting controversies, twists and turns, misunderstandings, admiration, humility or emboldened presumption, friendly dialogues or discourteous polemics follow one another. In the process, we will follow the slow and difficult work of elucidating reality in relation to a few selected questions concerning what the Genevan naturalist Charles Bonnet called "the great & murky matter of the generation of living beings."

Going back to the past to understand how what we know has been developed little by little, or sometimes by leaps and bounds, the first thing that strikes you is the constant presence of various obstacles to the progress of knowledge. Apart from the force of habit and the difficulty of asking the questions afresh, let's look briefly at four more of these obstacles in an attempt to grasp, at least roughly, the context in which the various speculations that have arisen over the years have emerged – or, on the contrary, remained permanent.

The first of these obstacles stems from the fact that animal societies, hymenoptera and the honey bees in particular, have been constantly used since antiquity in the Western world as a powerful metaphor and justification for the hierarchical organisation of human societies and the political and social relationships that exist within them.[1] This use of bestiary, which varied in space and time, had a moral and, in any case, political intention, with the aim of submission to the altar or the throne, and often to an alliance of the two. As a result, phenomena of no moral or political interest, or of a contrary nature, were left in the dark. A second obstacle stems from cultural prejudices, which are also specific to each era and each community, and which hinder the intellectual process.

As a result, if a few lights managed to illuminate the blind spots of knowledge, it was not easy to account for observations or speculations that did not fit into the traditional explanatory framework and clashed with an ordered but oriented vision of the world: how, for example, could a female be seen and accepted where all the political, moral and learned authorities saw a male and put the latter in the place to which biology, subject to the laws of evolution, had assigned him? How could it be argued that a mating between a queen, mistaken for a king, and a drone absent from the hives for part of the year was necessary for generation, when the latter's masculinity was sometimes denied?

Finally, two other obstacles, one relating to the lack of adequate means of observation, the other to methods of investigation, also hampered the perception and understanding of behavioural or biological phenomena. Time was needed for successive advances to be made in these two areas.

Not that the authors of past eras necessarily lacked intellectual probity, but in order to leapfrog over what could not or would not be seen, and in order for the animal model to continue to operate, they would resort, as we shall see in the chapters that follow, to theoretical constructions, some ingenious or more imprudent, others elegant or more improbable. These constructions were sometimes close to what we hold to be true today, or as far from it as they could get.

From the 17th century onwards, the growth of learned societies in France and Europe[2] and, above all, the many exchanges that developed between them helped to bring about times when the accumulation of observations and speculations was such that traditional visions were bound to give way. And so, not without resistance and

---

1   Other uses have been made, which have not always presented the bee as a virtuous example, whether, for example, to depict the clergy and the papacy in a disparaging and mocking way (WOOLFSON, 2010) or to express an economic theory – see "The Fable of the Bees: or, Private Vices, Publick Benefits" by the Anglo-Dutch economist Bernard Mandeville (1705).

2   See, for example, "Les circulations et les transferts agronomiques entre la Société économique de Berne et les sociétés d'agriculture françaises» (1757-1773) (ROBADEY, 2021).

a curious step backwards or sideways, the charm of metaphor gradually ceased to operate, making it easier to understand phenomena, helped by the changes taking place in societies themselves and by methodological and material progress.

In the following pages, the party has been taken to build a framework based on some of the great names in the history of the development of knowledge on our subject, and to include less illustrious or lesser-known contributors. In doing so, we cannot rule out the possibility that knowledge may have been developed at different times and in different places by keen observers, scholars or simple practitioners, but that we are not aware of it, that they have not written or that we have not read them. And if the pages that follow mention "discoveries", it is important to understand that, as François Huber, the Swiss naturalist whom we will meet frequently in this text, says: "practice has always preceded theory," and that those who think never do so ex nihilo.

Along the way, we may smile at some of the statements made, but sarcasm and contemporary interpretations are not appropriate for those of us who are arriving comfortably after the phenomena we are about to discuss have, for the most part, been explained.

> "If we had lived in their century, we would have dreamt like them, and they would reason like us, or perhaps better than us, if they lived in ours. We should consider ourselves fortunate to have been born at a time when reason has succeeded in destroying so many prejudices, and when it has shown us the certain paths we must follow to discover its truth."
>
> Réaumur, "Mémoire pour servir à l'histoire des insectes," 1740

Let's begin this detour into the past with the question debated for several centuries about the gender of the hive's inhabitants.

A history of the discoveries about the sexuality of the honey bee

Chapter I
A question of gender

Fig. 2 "Traité curieux des mouches à miel ..." *(A curious treatise on honey flies ...)* –
Anonymous author (1740). The engraving shows, according to the author's views,
6 types of bees: the King (A), small honey bee (B), medium honey bee (C),
large honey bee (D), grey honey bee (E), drone (F).

"In a word, all the combinations that can be made in relation to the sex and non-sex of the three kinds of bee have been, and there has been one adopted and given as true by some author."

Réaumur, "Mémoire pour servir à l'histoire des insectes," 1740.

From antique speculation to 17th-century science

As the French naturalist and physicist René Antoine Ferchault de Réaumur wrote, these are just some of the hypotheses that could have been formulated, which we will discuss in this first chapter in order to follow the tortuous path of the progress of knowledge on the question of the gender of the individuals populating the hive.

And to do this, as is often the case, we start with Aristotle. In Book V of his treatise *History of Animals* (343 B.C.) and in Chapter IX of Book III of his treatise *Generation of Animals* (322 B.C.), the philosopher examines the many theses that had been formulated up to that point.

Without going into detail about them, it should be noted that in his most recent treatise, the philosopher examines and then rejects them and, himself caught up in the prejudices of his time which made inconceivable that a female could be at the head of a hive and that the workers, also female, should have a stinger[3], formulates his own hypothesis, consisting first of placing not one, but kings or chiefs at the head of the hive. Then, if two other castes are clearly distinguished – the workers (whom the philosopher simply calls bees) and the drones – he comes to maintain that the kings or chiefs beget themselves and the workers, who in turn beget the drones, who are denied the slightest role in reproduction.

If the drone is not given a role, it is because, for Aristotle, it is not male. As the German zoologist Rudolf Leuckart wrote in the 19th century, their masculinity was long denied, because "it seemed quite incredible that the male sex, which in the human race and in the higher orders of animals occupies the privileged and leading position, could in any case be reduced to such a subordinate condition as that occupied by drones in the economy of the hive." (LEUCKART, 1861).

Finally, there was a difficulty to be resolved with regard to the worker bees, since Aristotle said that they were capable of engendering, and therefore female, but that they bore a stinger, a clear sign, he thought, of their masculinity. Here is how the philosopher attempted to resolve this difficulty, which was due as much to conceptions as to the limited means of observation available at the time: "Another hypothesis is that bees beget drones without mating, being indeed female in the sense that they beget, but having, like plants, the female and the male enclosed within themselves; which also means that they have fighting instruments."

Other ancient authors such as Virgil, Varron, Columella and Pliny took up Aristotle's theses to a greater or lesser extent. Thus, until the 17th century, the idea that a king was at the head of the beehive dominated, even if, at all times and as far back as Aristotle's time, some people saw things correctly. Many authors played on the political and/or religious metaphor with verve, praising the great qualities of the supposed king (of the hive) and, for some, the real king (the sovereign of their country). They include the Frenchmen Charles Estienne ("L'Agriculture et Maison rustique," 1564 and 1598), Olivier de Serres ("Théâtre d'Agriculture et Mesnage des champs," 1600) and the Englishmen Thomas Hill ("A Pleasant instruction of the perfect ordering of Bees," 1568) and William Lawson ("The Country House-wives Garden," 1618).

The idea persisted at least until Jan Swammerdam demonstrated in 1669[4]

---

3   "Nature gives no female weapons for combat," said the philosopher.

4   The French version of "Historia insectorum generalis ofte algemeene verhandeling van de bloedeloose dierkens" was published in 1682.

(SWAMMERDAM, 1682) that the queen is a female and the drone a male. Swammerdam placed the female worker in a third caste, but was mistaken about her sex, as we shall see in the next chapter. As for the question of the respective roles of the two sexes in procreation, it remains the subject of many diverse and often untenable hypotheses. We will also address this question in the next chapter.

It is well known that Jan Swammerdam demonstrated that the central individual in a colony of bees, which was still generally considered to be a male, was a female. What is less well known is that the Dutch naturalist and physician had a number of forerunners in this field. let's talk about the ones we know about.

In 1586, Luis Méndez de Torres, a Spaniard about whom little is known,[5] states in the first chapter of his "Tractado breue de la cultiuaciō y cura de las colmenas"[6] (MÉNDEZ DE TORRES, 1586) that the head of the bees is a female and that, laying eggs, she begets the three kinds of bees, queens, drones and workers: "You should know that the bee called Maesa, or Maestra, without the help of a male and without pain, sheds a semen of her own accord from which three kinds of bees are born: Maestras, drones and bees." He adds: "Thus, the semen being one and the same, it is through the diversity of the receptacles [the cells of the comb] in which it is placed that the three aforementioned genera are formed." Curiously, the book remained unknown and was not rediscovered until late in the 20th century.

Subsequently, in a work published in 1597, "Van De Byer" (CLUTIUM, 1597), a Dutch apothecary and herbalist from Delft, Dirck Outgaertsz Cluyt, also known by his Latin name Theodorum Augerius Clutium, also claimed that the king was a queen.

A few years later, in 1609, Charles Butler, an English naturalist and grammarian, published *The Feminine Monarchy* (BUTLER, 1609) – see Figure 3. The book can be considered the first scientific work in the field of beekeeping.[7] Butler wrote that the king was a queen and that she begat both workers and drones, which were also male. Butler accepted these facts as reported by observation ("reason and sense agree"). It is true that by this time, women had already sat on the English throne. However, even with a queen at its head, the beehive as a political metaphor lives on, and the idea that the social order as it was intended to be embodied by the colony of bees remains inscribed in the divine order. "Bees participate in divine reason and celestial

---

5   He is said to have been a resident of Alcalá de Henares (Spain) in 1586, a book mer-
     chant, a servant of the Duke of Alburquerque and a beekeeper in his native Sierra More-
     na region of Extremadura (GUZMÁN-ÁLAVAREZ, 2006).

6   "Brief treatise on the cultivation and care of beehives."

7   See the very interesting essay "Can Females Rule the Hive?" by Frédérick R. PRETE
     (PRETE, 1991).

influence," he wrote, no doubt assured of celestial support.

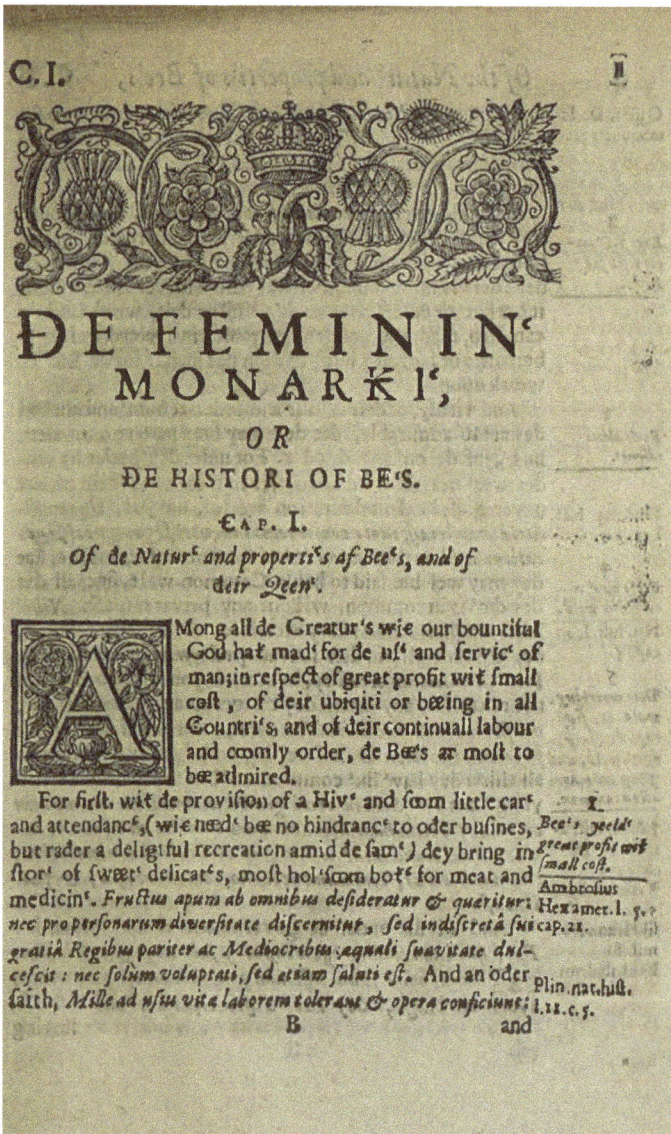

Fig. 3 Incipit of *The Feminine Monarchy* by Charles Butler (2nd edition of 1634).

Finally, to leave Swammerdam's main precursors there, let's say a word about the Englishman Richard Remnant, a prosperous Puritan merchant, who also asserted in 1637 that the beehive was a female monarchy governed by a queen (REMNANT, 1637). No doubt inspired by the Bible, he involved the divine creative breath in the matter:

"The females blow their brood into the cells or holes of their combs." For the author, there is no doubt that it is the females who produce the two castes – male and female – according to the needs of the colony, since they do so even in the absence of the males. The semen they blow is mixed with a substance collected from the flowers. The contents of the cells thus provided will be brought to maturity thanks to the heat produced by themselves and the males whose present use it is.

As far as the queens are concerned, the author, although not certain, thinks that they themselves give birth in the same way as the workers, i.e. by blowing, but in this case the semen is received from the males (whom he calls "drones"[8]) during the summer, although he admits that he does not know how. Nevertheless, he ventured the hypothesis of mating, noting that the male had an organ specific to reproduction and considering the anatomy of the queen.

Remnant, a fruit of his time, after conceding that there is a queen at the head of the colony, gives the lie a little further on:

> A cocke that's silent, and an hen that crowes.
>
> I know not which live more unnaturall lives,
>
> Obeying husbands, or commanding wives.

Be that as it may, Swammerdam was the first to base his assertions on solid anatomical evidence. And it was the microscope – a recent invention perfected by the Dutchman Antoni van Leeuwenhoek – of which he was a talented user, that enabled him to examine the anatomy, tissues and organs of bees on the basis of dissections that he carried out himself. We owe him the first representation of the reproductive organs of a queen and a male bee (Figure 5). The microscope and skilful scientists such as Jan Swammerdam, Antoni van Leeuwenhoek and Marcello Malpighi were responsible for major advances in our knowledge of insects (among other things) from the 17th century onward. These advances have left us with some magnificently illustrated works.

And since a microscope required lenses, perhaps the scientist knew the philosopher Baruch Spinoza who, in order to preserve his independence, worked as a lens polisher in Amsterdam, a trade for which he had acquired a great reputation.

---

8   In the 1520s, the term took on the figurative meaning of idler, lazy worker. From 1946 onward, the term was also used to designate unmanned flying machines.

Latecomers

It would take several decades before Swammerdam's theses were widely disseminated and then just as widely accepted. In the meantime, old theses will persist; and new ones will probably be imagined. Sometimes, even when they are known, Swammerdam's facts will be rejected.

Seeing double ...

The illustration at the beginning of this chapter (Figure 2) refers to a late 18th-century speculation. In this work, sometimes attributed to Talonné de la Bretonnière or Louis Liger, the author distinguishes six types of bees, whereas, as we have just seen, the existence of only three castes had been demonstrated more than half a century earlier.

A philosophical lag ...

Not far from a century after Swammerdam's "Historia insectorum generalis," Voltaire, an early French writer, philosopher and later beekeeper, seemed to reject both the idea of a queen at the head of a colony and its supposedly unbridled morals: "I don't know who first said that bees had a king. It probably wasn't a republican who first got the idea.[9] I don't know who then gave them a queen instead of a king, nor who was the first to suppose that this queen was a Messalina,[10] who had a prodigious seraglio, who spent her life making love and giving birth" (VOLTAIRE, 1789). We can smile to hear the staunch defender of freedom of conscience, who coined the superb phrase: "Everyone goes to heaven by the path he chooses,"[11] refusing to let the bee soar to her celestial love affairs, perhaps with a touch of moral disapproval.

In reality, the philosopher, who had observed only a few bees and did not write only admirable things, endorsed the thesis of Jean-Baptiste Simon, lawyer at the Paris Parliament, royal censor and beekeeper, who maintained that each of the three castes begat itself: a royal couple reigns over the subjects of the hive and perpetuates "this royal race"; drones are of both sexes, just like common bees, and each of these castes reproduces on its own to form the body of the colony (SIMON, 1742). This fabulous thesis stems from the fact that the beekeeper does not want to admit, on the one hand, that a male is devoid of a stinger, on the other hand, that an individual of one caste can engender an individual of another caste and, finally, that a queen alone founds the colony.

---

9   Voltaire was in favour of a liberal and enlightened monarchy.

10   An empress of ancient Rome who, as history has it, had a great appetite for sex.

11   A superb sentence, albeit incomplete, because in his writings Voltaire does not defend those who do not believe in heaven, quite the contrary.

Engravings based on microscopic views

An early representation of a microscopic view was drawn by Francesco Stelluti in 1625 (Figure 4) and published by Federico Cesi (see also Figure 1). However, as the academic Matthew Cobb writes: "However, Swammerdam was the first to make in-depth studies of social insects, using a combination of careful observation, dissection and experimentation. As such, he played an important role in laying the foundations of modern biology in terms of both his results and the methods he used." (COBB, 2002).

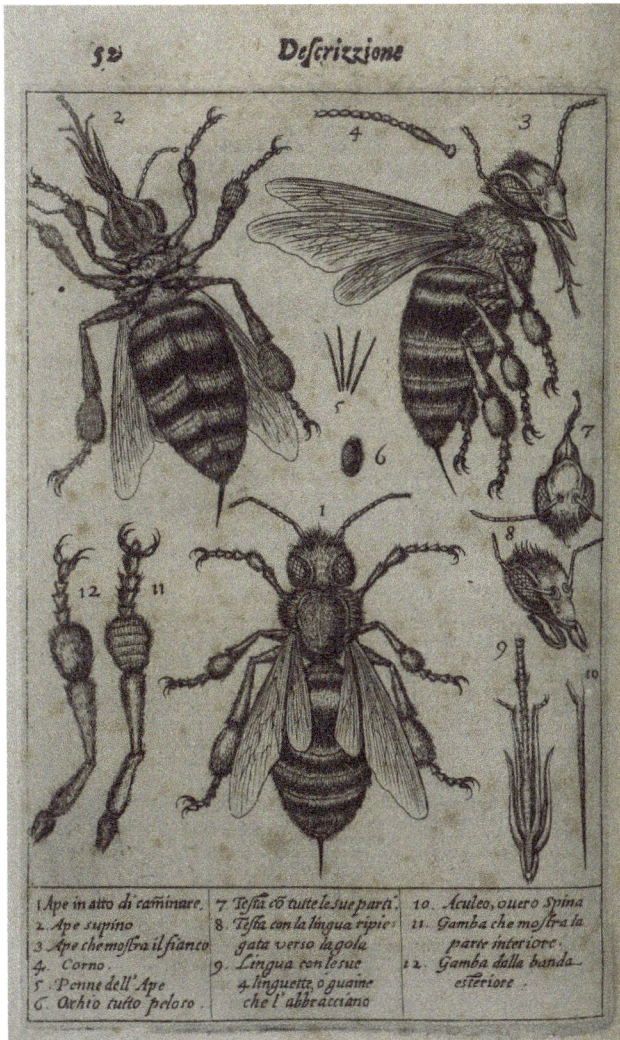

Fig. 4 « Three bees » by Francesco Stelluti, 1625.
(Deutsche Digitale Bibliothek - Photo under Creative Commons BY-SA 4.0)
First publication of drawings based on microscopic views from the book "Apiarium."

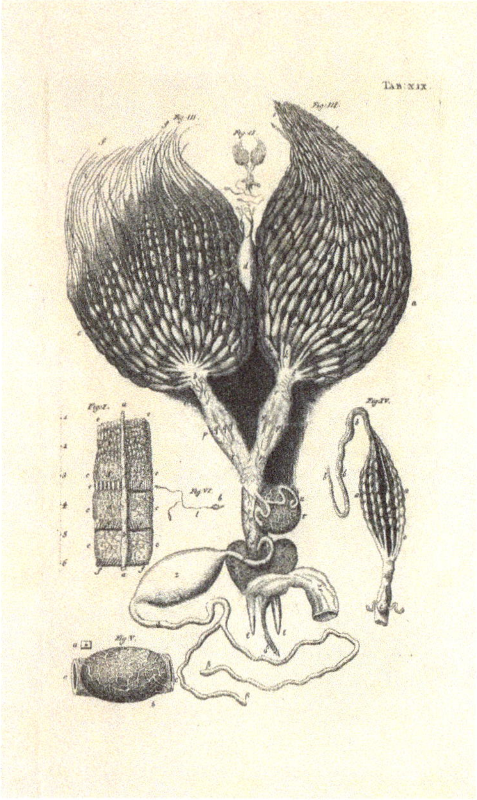

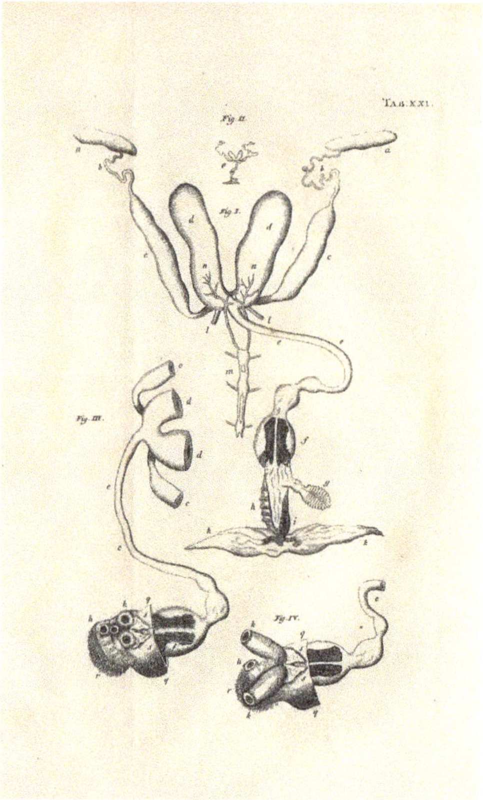

Fig. 5 Drawing of the ovaries of a queen (Table XIX, on the left) and drawing of the male organs (Table XXI on the right) seen under a microscope, taken from the work "Bybel der Natuure" by Jan Swammerdam (1737).

Chapter II
How are babies made?
The first hypotheses

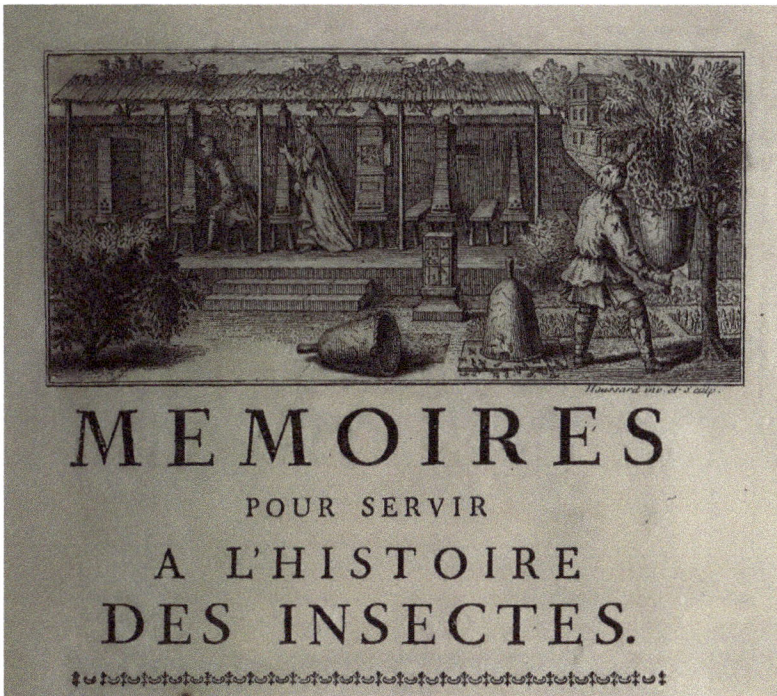

Fig. 6 Illustration of the introduction to volume V of "Mémoire pour servir à l'histoire des insectes" (*A memoir of the history of insects*) by René Antoine Ferchault de Réaumur, 1740. The engraving shows the observation hives used by the naturalist and the capture of a swarm.

In the previous chapter, the first stage of our journey into the past, we saw how the question of the gender of the hive's inhabitants was approached. By the end of the 17th century, it had been established that a bee colony was made up of individuals belonging to three castes: a queen — a female capable of begetting — drones whose masculinity was finally recognised and, finally, common bees whose sex was uncertain and who did not beget.

As a result, like most species, bees are subject to a sexual mode of reproduction involving, but this was not known until the 19th century, a process that implies the production of male and female sex cells – gametes – by cell division (meiosis) and then by the fusion of these gametes (fertilisation), resulting in a diploid individual possessing equal amounts of the genes of its two parents. We will see later that bees can also reproduce in another way.

In the meantime, one question remains: how do these individuals reproduce? As we shall see, there are several hypotheses, depending on the period and the author, from the virginal bee freed from the guilty and sinful pleasures of copulation to the simple, and ultimately most innocent, recognition of mating.

Let's go back to Aristotle for a moment. According to him, only kings and bees beget. They do beget, but without mating! The philosopher's hypothesis on this subject is unclear, but he seems to opt for reproduction "like [that of] some species of fish," without ruling out spontaneous generation, which means that the appearance of certain beings proceeds from inanimate or even decomposing matter. In Chapter VIII of Book III of his treatise "Generation of Animals," Aristotle writes: "There are insects that come from copulation, as do birds, viviparous animals and almost all fish; others are born spontaneously, like some plants."

The theory of spontaneous generation, which can be found, for example, in the form of the bugonia ritual in the myth of Aristaeus, was opposed by Jan Swammerdam and invalidated on several occasions, only to receive the *coup de grâce* with the work of Louis Pasteur in the 19th century. It had been a long time since anyone had attributed this mode of generation to bees.

We know that Swammerdam, like Méndez de Torres mentioned above, doubted the possibility of mating between the queen and the drone: "Firstly, although the males of bees, among the parts that characterise their sex, seem to have one that corresponds to the rod of large animals, this part is nevertheless not sufficient for true mating, not only because it has no use for the seminal liquor, but also because its shape and situation would not allow it to enter the body of the female." (SWAMMERDAM, 1737).

However, Swammerdam differs from his predecessor by recognising the male's role in procreation: "The male has no other care than to fertilise [the] eggs even before they are laid, & he fertilises them for a whole year." According to this theory, fertilisation must therefore be repeated before a swarm leaves with the previous queen. It was no doubt inspired by Aristotle's distant analogy with certain species of fish that spread their seed over the eggs laid by the female, and because he had noticed a strong odour given off by males when they are present in large numbers, that he conjectures that "the subtle vapour which rises from the seminal liquor [*aura seminalis*] of the males spreads throughout the basket, penetrates the body of the female, & fertilises the three kinds of eggs which are contained in her ovary."

We note that the Dutchman is cautious about making any peremptory assertions, because he concludes as follows: "until we have reliable experiments that can tell us the true manner in which bees are fertilised, experiments that I do not believe to be impossible, I will stick to my feelings." It should also be noted that the scientist admits the possibility of a refutation of his positions. What's more, there was a genuine scientific approach at work, although divine explanation, another type of discourse, was not absent from his writings to such an extent that at the height of his mystical period, some of his texts would be cluttered with considerations imbued with extreme religiosity. Under the influence of a fanatical devotee, he interrupted his work, these "amusements of Satan," for several years and tried to burn his notes and drawings. Freed from her influence, he resumed his work and wrote his "Bybel der Natuur" (*Bible of Nature*), which was not published until 1737, more than 50 years after his death in 1680. The work, written in Latin and Dutch, was partially translated into French and English.

On the subject of workers, Swammerdam concluded that, since he was unable to observe the ovaries of those he dissected, they "contribute nothing to the propagation of the species," as they are "neither male nor female" and "incapable of generation." Further on, noting the great attraction that the queen exerts over them, he assumes that they are "natural eunuchs."

Subsequently, between 1734 and 1742, René Antoine Ferchault de Réaumur, a French physicist and naturalist, published his voluminous "Mémoire pour servir à l'histoire des insectes" (*A memoir of the history of insects*) in seven volumes. In the 9th memoir of the 5th volume, published in 1740 (RÉAUMUR, 1740), the naturalist returned to the writings of Swammerdam, whose arguments on the question of how eggs are "vivified," i.e., fertilised, he examined very seriously and finally rejected: "There is therefore every reason to believe that the fertilisation of the mother bee is not carried out in the extraordinary way that Swammerdam believed it was." Although Réaumur immediately ruled out the possibility of mating taking place outside the hive – "there is no reason to think that she leaves the interior of the hive where she so likes to keep herself," he wrote – he remained convinced that mating did take place, by analogy with wasps whose mating he reported having seen.

Réaumur mentions the existence of the ejaculatory orifice of the male organ which the Dutchman had not observed, but, no more than Swammerdam,[12] he will not find the true function of the vesicle close to the ovaries. Mention should be made of Marcello Malpighi, whom Réaumur had read but did not take into account on this point. This Italian anatomist had highlighted the function of what would later be called the spermatheca[13]: "The inner cavity is filled with a certain mucous juice very similar to a barley tea; also, as it is in this cavity, as will be seen later, that the semen is ejaculated after the introduction of the penis, that it remains there and is completed there, I thought that this was the uterus where the semen mixes with this particular juice, to then go little by little, in all probability, through the duct and impregnate the eggs as they pass" (MALPIGHI, 1686).

Réaumur assumed that fertilisation of the eggs took place in the queen's body, seeing that she could lay eggs in summer after he had observed that the males had disappeared, and also at the end of winter when the males were not yet present. However, wondering about what would later be called embryology, he mentions the observations that led Leeuwenhoeck[14] to see small worms in the milky liquors of the males of different species of animals, but specifies that this "is not yet as proven as would be desirable."

Finally, Réaumur recounts, and it's rather amusing, his attempts to induce mating between a young queen and a drone. He mentions, not without sparing his readers' prudishness, the "enticements" that the young nubile queen puts the male through and her "immodest positions." The naturalist's attempt was unsuccessful, but this did not prevent him from maintaining the hypothesis that mating takes place in the hive, on the initiative of the mother, who "makes advances to the males she likes ... it is up to her to draw them out of their state of indolence and coldness." What he saw here as a "reversal of order" seemed necessary to him because, conversely, the ardour of a large number of males would leave the young female no rest. On the other hand, the incestuous aspect resulting from the *intra coitum* hypothesis, if it could not be raised at that time with genetic considerations (the evils of endogamy), could not be raised either in the name of morality to which, nevertheless, there was no lack of

---

12   Réaumur states that Swammerdam considered it to be a pulmonary bladder designed to supply air to the ovaries. In fact, Swammerdam located in this cavity the production of "the viscous liquor that holds them [the eggs] by one of their ends against the bottom of an alveolus'".

13   The French naturalist Jean-Victor Audouin (1797-1841) is credited with coining the name. The spermatheca was also known as *Audouin's copulatory pouch.*

14   Antoni van Leeuwenhoek (1632-1723), Dutch scientist, is a precursor of cell biology and microbiology, continuing the work of his compatriot Jan Swammerdam. He communicated his discovery to the Royal Society of London in 1678.

desire to subjugate men and beasts alike.

François Huber would later repeat Réaumur's attempt to provoke mating, without any more success, before considering the question from another angle, as we shall soon see.

Figure 6 inserted at the beginning of this chapter and Figure 7 below illustrate the observation hives used by Réaumur.

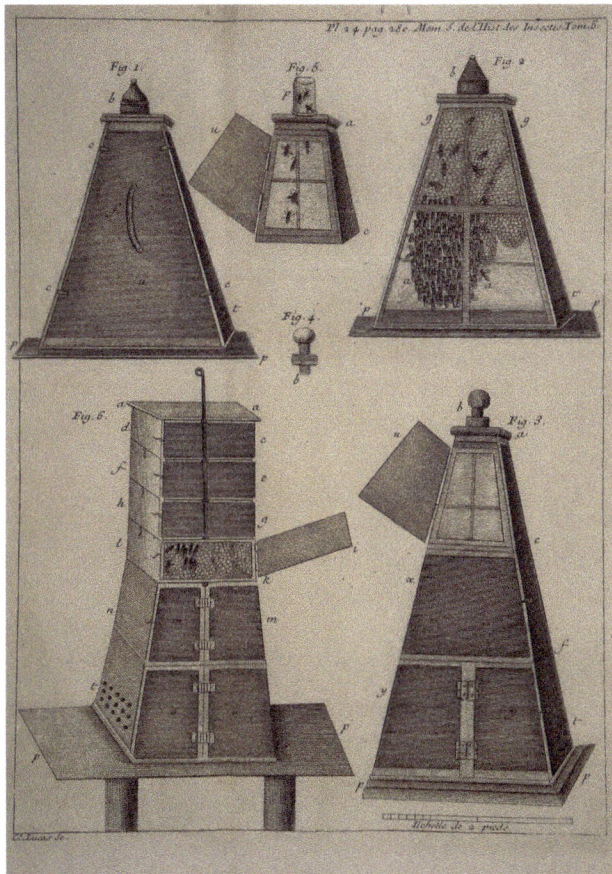

Fig. 7 Three different types observation hives used by Réaumur (RÉAUMUR, 1740)

A few years later, in 1770, the German pastor Adam Schirach published his book "Ausführliche Erläuterung der unschätzbahren Kunst, junge Bienenschwärme, oder Ableger zu erzielen: nebst einer natürlichen Geschichte der Bienenkönigin" (*Detailed explanation of the invaluable art of raising young swarms or offspring: together with a natural history of the production of the queen*; SCHIRACH, 1770), translated into French the following year (SCHIRACH, 1771). After a long period of reflection and based on the experiments of one of his friends, Mr Hattorf, a report on whom he included in his work, Schirach concluded that males did not intervene in procreation: "What is singular about this Queen is that she is fertile & produces vivified eggs without having known a male." In this regard, he cites his correspondence with Charles Bonnet, a naturalist and philosopher from Geneva who was a pupil of Réaumur, and who demonstrated parthenogenesis in aphids. The thesis of an absence of copulation is therefore not incongruous for our naturalist. Schirach will also highlight a major fact that we will discuss in the next chapter.

At the same time, other speculations on fertilisation were being made. One of them, supported by the Englishman John Debraw, postulated that males deposited their semen on each of the eggs laid by inserting their posterior into the cell (DEBRAW, 1778). This is the impregnation theory. This apothecary claims to have observed this behaviour on several occasions and to have noted a clear difference between the product of the supposed semen and royal jelly. Of course, as François Huber notes in the first letter of his "Nouvelles observations" (HUBER, 1792), the author overlooked, among other factors, the fact that brood was reared even in the absence of drones, at the beginning and end of the season. But for Debraw, two types of drones exist and one of them is not expelled from the hive and spends the winter there. However, Huber refuted Debraw's thesis by observing the development of brood in the absence of any males in the hive.

Since François Huber's name has been mentioned several times so far and we will come across him again later, let us digress here to point out that the Swiss naturalist, who became completely blind at the age of 22, undertook all his observations mainly with the help of François Burnens, his servant who became his devoted and faithful assistant, and his no less devoted and faithful wife, Marie-Aimée. The observations mentioned in this text are those undertaken by the naturalist assisted in this way.

Having reached this point, let us leave aside for a moment the question of the procreation of individuals in the hive, which the authors we have just mentioned have not completely exhausted, just long enough to mention other important advances that were made at the same time.

Chapter III
To each his own egg?

Let's stop for a moment to consider the question of how many types of eggs are needed to produce the three castes of bees.

For Swammerdam, as we have seen, there are three types of eggs, one for each of the three castes. These eggs are conceived and fertilised in the queen's body. Réaumur said the same thing, although he did not share the Dutchman's views on the way in which the queen is fertilised, as we have also seen.

Adam Schirach, whom we have already met (see Chapter II), looking for a way of multiplying swarms, found that by splitting a colony in two, the workers would stretch out a royal cell in the part that did not receive the queen, provided it contained young larvae. He published his method of creating artificial swarms in a treatise (SCHIRACH, 1761) − see Figure 8. This discovery, which was of great interest both practically and theoretically, won him great renown in the provinces and states of the Holy Roman Empire, as far afield as Poland and Russia (the Empress sent an envoy for him), but also earned him a great deal of criticism, of which he complained: "How many people condemn any novelty without examination! How many people reject at first glance anything that requires a little more attention and work than they are accustomed to giving themselves!" Critics could not accept that a queen could be produced from an egg that could just as easily have produced a worker.

Royal egg or common egg?

In his book published in 1770 (see Chapter II), the German pastor specifies what he believes to be the determining cause of the differentiation between a queen and a worker from the same egg: "The queen bee is formed from one of the three-day-old worms of the worker bees. It is to this three-day worm that the workers in the hive, where it is found, make a kind of special nest, which looks very much like an oak acorn. It is called the Royal Cell. It is in this Royal Cell that the Bees carry a particular kind of food, quite different from that which must be used for the sustenance of other Worms." He notes that bees do not produce a royal cell from older larvae because, he says, "in my opinion, the latter [worms over three days old] have already reached

19

a stage where food can no longer produce the required effect; instead, in the others [3-day-old larvae], as the parts are not yet well developed, this particular food can produce its effect: For I firmly believe that it is only by means of the juices contained in this particular type of food that the queen's feminine parts develop."

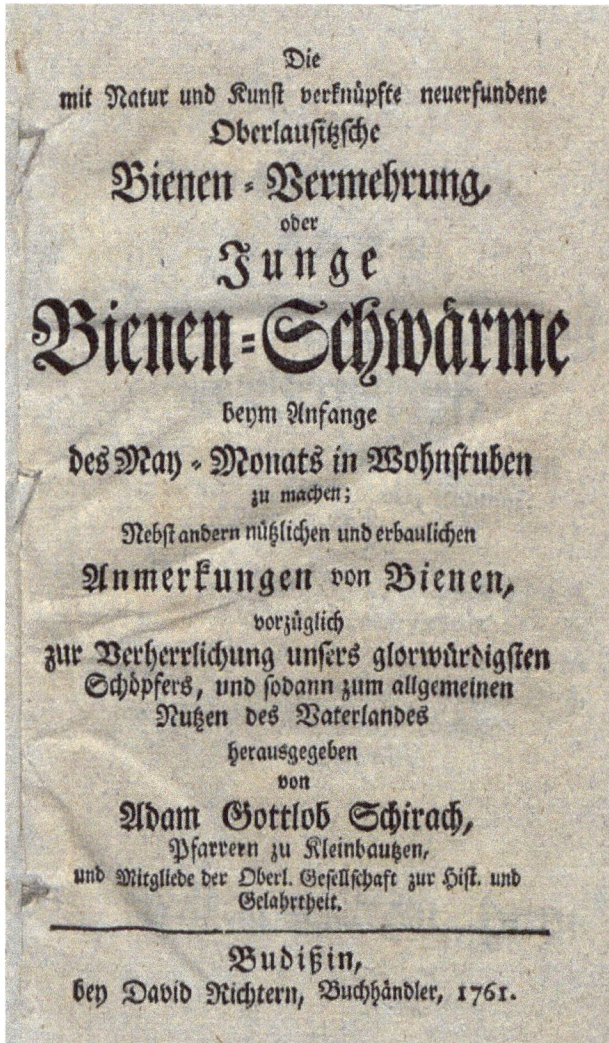

Die
mit Natur und Kunst verknüpfte neuerfundene
Oberlausitzsche
Bienen = Vermehrung,
oder
Junge
Bienen=Schwärme
beym Anfange
des May = Monats in Wohnstuben
zu machen;
Nebst andern nützlichen und erbaulichen
Anmerkungen von Bienen,
vorzüglich
zur Verherrlichung unsers glorwürdigsten
Schöpfers, und sodann zum allgemeinen
Nußen des Vaterlandes
herausgegeben
von
Adam Gottlob Schirach,
Pfarrern zu Kleinbautzen,
und Mitgliede der Oberl. Gesellschaft zur Hist. und
Gelahrtheit.

Budißin,
bey David Richtern, Buchhändler, 1761.

Fig. 8 Title page of Adam Schirach's 1761 work, "Die mit Natur und Kunst verknüpfte neuerfundene Oberlausitzsche Bienen-Vermehrung" (The New Reproduction of Upper Lusatian Bees, Linked to Nature and Art, or Young Swarms of Bees)

According to his views, these eggs are produced by one of the two specialised branches that make up the queen's ovary, the other producing eggs that give rise to males whose only use, he says, is brooding.

The pastor reported this observation to the Swiss naturalist Charles Bonnet, who did not give him spontaneous credit, but in turn sought the advice of his compatriot François Huber. The latter, although astonished that the same egg could produce two individuals as dissimilar as a queen "capable of prodigious fecundity, but unfit for work of any kind" and a worker "sterile, but capable of the most astonishing industry," nevertheless did not admit "that worker bees are monsters or imperfect individuals: too many precious gifts, too much industry, too much activity, have been bestowed upon them; too many marvels result from their instinct and their organisation for me to consider them as the scum of the species or as imperfect beings, relative to queens."

François Huber repeated the experiment and confirmed Schirach's discovery. He even determined that one- or two-day-old larvae could also lead to royal breeding.

Indirectly, Schirach's discovery was tantamount to attributing the female gender to workers, and this was another advance, since Swammerdam himself had considered workers to be "incapable of generation" and Réaumur concluded "that they are neither male nor female, that they are absolutely devoid of sex." Perspicacious, Schirach presumed that the ovary of the worker did exist, but that it "must be very small in these flies since it escaped the piercing eyes of the illustrious academician [Réaumur]." Thus, for Adam Schirach, there are only two kinds of eggs and not three, which seems to be consistent with the fact that, according to him and as we have seen above, the ovary is made up of two branches. The definitive proof of this will come a little later.

Adam Schirach was right, both in terms of the possibility of rearing a queen from the same egg, which could just as easily have produced a worker, and in terms of the differentiating effect of the special food that would later be called royal jelly.

Egg-laying workers?

Proof of the femininity of workers

At that time, Johann Riem of the Economic Society of Lautern (in the Palatinate, a state of the Holy Roman Empire), seeking to refute Schirach's thesis by arguing that the workers had transported, without Schirach's knowledge, a queen egg to where the pastor found a royal rearing which he mistakenly believed to be undertaken from a common worker larva, reported the observation that common workers lay eggs under certain circumstances.

In truth, this last fact did not constitute a discovery, as the phenomenon must not have escaped the notice of any beekeeper with the slightest attention to detail. In fact, the phenomenon has been known since at least the time of Aristotle. This is what the philosopher says in his first treatise (see Chapter I), when he wants to demonstrate that chiefs give rise to bees, but that it is only bees that give rise to drones: "What proves it, as they say, is that there can be generation of drones without there being chiefs and that there is no generation of bees."

Adam Schirach, whose thesis withstood Riem's attempt at refutation, also noted these egg-laying workers, and drew the obvious additional conclusion as to the femininity of the workers.

For his part, François Huber went even further: He *proved* the femininity of the worker. The 5th letter of the "Nouvelles Observations" (HUBER, 1792), dated 25 August 1791, relates the observations made in the summer of 1788, the purpose of which was to study the case of egg-laying workers, about which Johann Riem had written some time earlier. The naturalist and his assistant François Burnens found the proof by dissecting the first bee caught laying eggs. For a while, they suspected that the egg-laying bees were "little queens" and not common worker bees. Although they were not very skilled at this operation, they were able to distinguish in bees that were identical in every way to workers "smaller, more fragile ovaries, composed of fewer oviductus than the ovaries of queens; the nets containing the eggs were extremely fine, with slight swellings placed at equal distances. We counted eleven eggs of appreciable size, some of which seemed ready to be laid."

Dissections were carried out again in 1811 by much more expert hands, when the naturalist called on Christine Jurine (1776-1812), a Swiss illustrator and naturalist, who, at his side, dissected both egg-laying and non-egg-laying workers and also noted the presence of ovaries similar to those of the queen, but incompletely developed. These last observations were published by François Huber in Volume 2 of the 1814 edition of his "Nouvelles observations sur les abeilles" (HUBER, 1814) and are accompanied by an illustration in the young naturalist's hand – see Figure 9. In the words of François Huber: "Thus vanishes that old theory which presented neutrals in bees; and the organisation of these flies which have so excited our admiration, by being related to universal laws, presents one of the most remarkable physiological phenomena."

Karl Theodor Ernst von Siebold, Bavarian anatomist and zoologist, demonstrated in 1843 that the spermatheca of workers also exists in a rudimentary state (DARESTE, 1858).

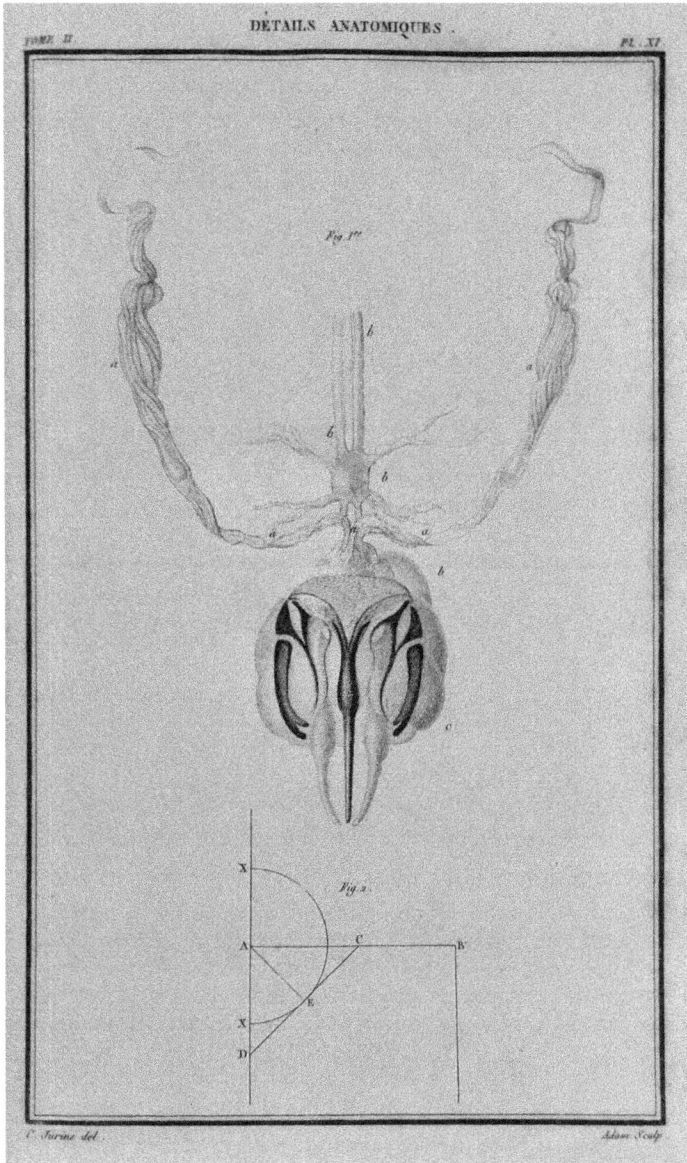

Fig. 9 Ovaries of a worker bee (top). Dissection and illustration by Christine Jurine. (HUBER, 1814)

However, it remained to be explained why the workers were laying eggs. François Huber, without any certainty but wishing to provide food for thought, ventured the hypothesis that laying workers were those whose larvae had developed close to a royal cell and had been able to benefit from the same royal food, albeit in quantities too small to allow full development, but sufficient to induce egg-laying in certain circumstances that remain unclear. Claude Philibert Lombard, the famous Parisian beekeeping teacher, supported this thesis (LOMBARD, 1805).

It was not until much later that it was shown that it was the absence or weakening of certain royal pheromones and their inhibiting effects on the ovarian development of the workers that caused some workers to lay eggs. As for the question of how the egg-laying workers are fertilised, Huber admitted that he couldn't explain it.

Finally, since the workers laid eggs and only males were produced, some went so far as to claim that the workers were exclusively responsible for generating males, thus partly supporting the Aristotelian thesis. In addition to more rigorous observations, it was, as Jan Dzierzon put it, the introduction of the Italian bee into northern Europe that definitively ruined this thesis: "When it was observed that a queen of the Italian race was introduced into a colony of the common [i.e., non-Italian] race, the males that appeared were also of the Italian race" (DZIERZON, 1882). The Italian bee is also the most elegant proof of a great discovery made by the Silesian pastor, which we will discuss in Chapter V.

When some move forward, others delay …

At the beginning of the 1880s, almost a century after François Huber, there were still opponents of this well-established fact that female workers lay eggs. These included the vehement Italian ecclesiastic Giotto Ulivi, whom we will meet in Chapter V, and the slightly less vehement Jean-Baptiste Leriche, a French beekeeper and founder of the *Société d'Apiculture de la Somme*. Leriche, responding to an article by Charles Dadant published in the Bulletin de la Société d'Apiculture de la Gironde in July 1884, criticised the thesis of the laying worker, which he said he had never observed. Like others before him, he could not accept that an individual of one caste could beget an individual of another caste (or only if it's a queen), nor that egg-laying could take place without prior mating. He gave us this fiction: "… that there are real fertilised mother bees that only produce drone eggs and that these can be so small that they can be mistaken for workers" (LERICHE, 1884).

Three in one?

As we have just seen, Adam Schirach and François Huber reduced the number of egg types used to generate individuals of the three bee castes from three to two: one for males and another for females.

Could one go further and assert that a single type of egg is sufficient and is the origin of either a male or a female? To be able to say this, the final mechanism of egg fertilisation still had to be determined. At the end of the 18th century, one had the idea that the male's semen played a role, but one was at pains to explain the real mechanism of what was then called the vivification or impregnation of eggs. Here is what François Huber had to say about it in 1791: "According to the opinion of a great physiologist, seminal liquor is only a particular stimulant which acts on the germs like a very substantial and very appropriate food for their development. Bonnet, in his attempt to explain Schirach's theory, made use of this hypothesis; he says somewhere: 'I have established, on the basis of evidence which seems to me to be solid, that seminal liquor is a real nourishing fluid and a stimulant, and I have shown how it can produce the greatest changes in the inner parts of embryos.'"

Until then, the theory of direct impregnation of all eggs, which François Huber believed took place fairly high up in the oviducts, prevailed despite Malpighi's observations about the copulatory vesicle – the spermatheca (see Chapter II).

The Dutchman Antoni van Leeuwenhoek identified animalcules in seminal fluid in the first half of the 17th century (see Chapter II). Nothing more was known at the time, but this did not prevent speculation. One of them was that one of these corpuscles penetrated the egg to develop and become the adult being, the egg being reduced only to its protective and nourishing functions. This was the theory of preformation, supported by the Church but later abandoned.[15] However, with these animalcules, a thread was held that would lead the knowledge a little further.

---

15   Nowadays we find a remanence of this idea in the form of the story that children are sometimes told about the little seed placed by Dad in Mom's belly.

Chapter IV
How are babies made? Coda

Fig. 10 Mating of a queen and a male *Apis mellifera*.
Drawing by Fany Eggers, courtesy of the author.

Until then, according to our authors (see Chapter II), either there is no copulation, or it does take place, but in the privacy of the hive.

According to Slovenian authors Andrej ŠALEHAR and Franc ŠIVIC (SALEHAR and SIVIC, 2021), it was Giovanni Scopoli, an Austrian doctor, naturalist and entomologist from the Italian-speaking Tyrol, settled for a time in the Duchy of Carniola (now Slovenia) who, in 1763, was the first to mention aerial fertilisation of the queen in his book "Entomologia Carniolica" (Figure 11) published in Latin (SCOPOLI, 1763). Describing some 1,153 species of arthropods and insects in the region, he says of the queen *Apis mellifica*: "She is surrounded by several males; she flies away and is impregnated in the air." Following Scopoli, other authors mentioned aerial fertilisation:

Peter Pavel Glavar in 1768, Matej Furlan in 1768 and 1771, Anton Humel in 1769, 1771 and 1773, Anton Janscha in 1771 and 1775, Adam Schirach in 1773, and François Huber in 1777 and 1778.

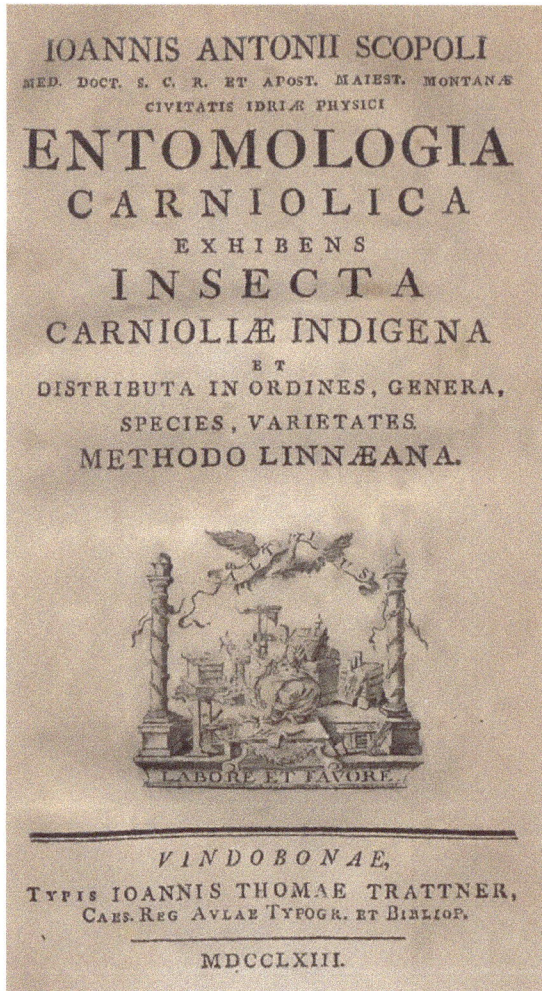

Fig. 11 Title page of Giovanni Scopoli's 1763 work.

On the other hand, while it is true that the observations reported by Janscha and his precursors on the fertilisation of queens in flight predate by a few years[16] those of François Huber, to whom this discovery is usually attributed, Huber was incomparably better at describing the phenomenon.

---

16   François HUBER indicates in his second letter to Charles BONNET that the observations
      reported in his previous letter date back to 1777 and 1778. JANSCHA's observations
      were published in 1775, two years after his death.

It was in his first letter to Charles Bonnet (HUBER, 1792) that the naturalist described his first observations, which date back to the summer of 1788. He scrutinised the departures and returns of queen fertilisation flights and, in this case, the presence of the fertilisation sign, which he described as follows: "The posterior part of her body was filled with a white, thick and hard substance; the inner edges of her vulva were covered with it; the vulva itself was half-open, and we could easily see that its inner capacity was filled with the same substance. This substance rather resembled the liquor with which the seminal vesicles of males are filled, & we found them perfectly similar to each other as regards colour & consistency." In his reply to this first letter, Charles Bonnet showed himself to be very enthusiastic and admiring of his compatriot's observations, and Huber found in this a recognition that his training in physics and chemistry and his blindness did not allow him to hope for at first, as he himself would say.

Huber soon added to his initial observations: "What we took to be a residue of the prolific liquor, was really the parts of the male's generation … the lenticular body, or lens." For the naturalist, this lenticular body is the penis, and it is the part that remains engaged in the sting chamber once the male has withdrawn. François Huber also notes that a queen can undertake several flights in succession, although he indicates that a single flight is sufficient most of the time. What's more, he notes that these matings render the queen fertile, if not for life, at least for two years (he had not yet observed more than that), and that they are never repeated. Previously, Anton Janscha, a Slovenian farmer and remarkable beekeeper, had observed that a queen could undertake several flights on the same day or overnight and that, after the age of 6 weeks, fertilisation was no longer possible (JANSCHA, 1775). Janscha also claims that the queen remains fertile for the rest of her life.

This is how Jacques Delille, a French poet, will sing what we owe respectively to Aristotle, who was the founder of the philosophical school of the Lyceum of Athens ("the teacher") and the tutor of Alexander the Great ("the conqueror of the world"), Réaumur and Huber:

The famous teacher of the world's conqueror

Wanted the bee to be fruitful without a husband;

And Réaumur, less jealous of her chastity

Prostituted their queen to many husbands:

At last, the learned custodian of their hymen,

The blind Huber saw it through the eyes of others,

And on this great problem a new day has dawned.

The queen, he tells us, on the day of the hymen,

Comes out of her new fires worried and astonished,

At the gates of the palace still hesitates for a long time;

At last her wing opens, she takes flight,

And far from mortal eyes, mysterious lover,

Carries into the air the love that torments her:

Her lover was watching her, and full of the same fires,

He leaves, flies, reaches her and enjoys, in the skies;

She soared as a virgin, she descended as a fertile mother.

«Les trois Règnes de la nature"
(The three kingdoms of nature), extract from the seventh Canto.
Jacques Delille, 1808

By the end of the 18th century, the mating of the queen in flight was a certainty. However, for a good century, the debate was vigorously, even arrogantly, maintained by a few deniers,[17] despite the accumulating evidence. Here are two testimonies that Lorenzo Lorraine Langstroth, an American pastor and beekeeper, reported in the pages of the French magazine L'Apiculteur in December 1861 (LANGSTROTH, 1861). The first dates back to 1849: "I saw twenty or thirty drones in rapid pursuit of a

---

17   For example, we can read debates of this type in the pages of the journal L'Apiculteur throughout the year 1881.

mother, at a height of twenty or thirty feet from the ground. The group occupied an apparent space of two feet in diameter: sometimes, in their course, they descended ten feet from the ground and then rose again, going from north to south. I could follow them for about 100 paces, after which a building made me lose sight of them. The group resembled fig. 3 and 4 [see Figure 16], except when it approached the ground: then the mass became more compact, and circular, − and formed such a thick whole that the insects seemed to touch each other as they flew, fig. 5 [see Figure 16]." This is what was later called a drone congregation area.

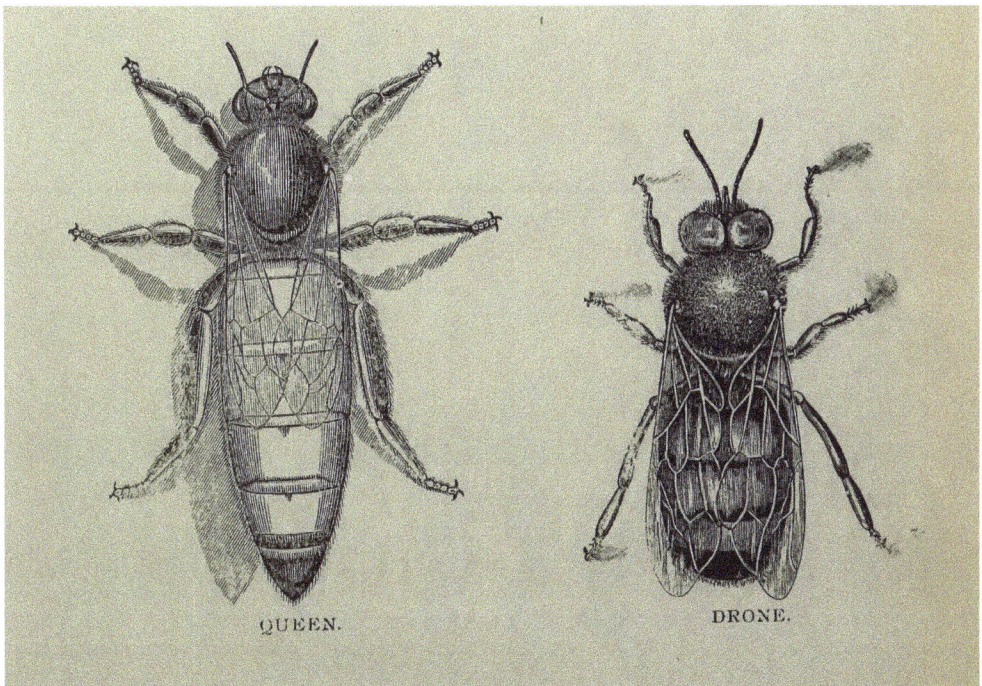

Fig. 12 A queen and a drone − Illustration from "The ABC and XYZ of Bee Culture" by Amos Root (1905).

The second testimony dates back to 1858 and relates a more direct observation of the Reverend M. Millette: "One of these mothers was seen flying, and a moment later was stopped by a drone. After having flown for a yard, they fell to the ground together, clinging to each other; the writer followed them at the very moment when they had just fallen; and, as the drone, having freed itself from the embrace, was about to set off again, the writer seized them both, carried them to the house, where he set them free : Then the mother flew to the closed window; the drone, after dragging itself in circles, fell to the floor at the bottom of the window and died after a few minutes."

In addition, as many beekeepers and naturalists persisted in denying the obvious, observations were made up until the early 20th century with young queens locked up, taking care to keep the males out of the hives and, with a few exceptions, we ended up with what François Huber had already observed a long time before: no mating under the duvet, in the house, but only outside, in flight! The exceptions mentioned above were, of course, either poorly conducted experiments or experiments designed to refute the evidence ...

Chapter V
Looking for the father

The parthenogenesis scandal

François Huber, a sagacious naturalist whom we have already met, noted in his 3rd letter of 1791 that he had first observed a queen laying eggs in June 1787 that produced only drones. He did not allow this queen to leave the hive until the 36th day after her emergence. She returned to the nest with the sign of mating and produced only males. The naturalist conducted a number of experiments, which showed that mating beyond the 20th day after emergence invariably produced this effect. However, he was unable to explain the phenomenon in terms of his thesis that all eggs are fertilised, which seemed to him to be the case since, although delayed, a mating had taken place each time. And if all eggs are fertilised, François Huber admits in his 5th letter, which deals with the question of egg-laying bees, that he does not know how they are fertilised.

In fact, as long as one did not assume that the eggs resulting in males were not fertilised, one has left the question unresolved.

The solution came as the middle of the 19th century approached, and with it, scandal.

Jan Dzierzon, a native of Silesia,[18] priest, beekeeper and naturalist, first published his theory of what would shortly afterwards be called parthenogenesis in November 1845 in the fledgling Bavarian periodical Bienen-Zeitung (DZIERZON, 1845) – see Figure 13 below. This theory, expressed simply, states that the drone develops from an unfertilised egg. It therefore has a mother but no father.

Jan Dzierzon sets out what he believes to be a theory that "allows us to make sense of all the most puzzling events in the hive." At the same time, he is aware that there will be no shortage of reactions.

In the world of beekeeping, we know all about the long and bitter battle in France and more widely in Europe between the partisans of the fixed-comb hive and the partisans of the movable-frame hive, but we know less about the battle between

---

18   Silesia is today included for the most part in what is now Poland.

the deniers of the theory and Jan Dzierzon and the supporters of parthenogenesis. And, as one would expect, the reception of these ideas caused tempers to flare, and reactions were violent, not only because they ran counter to the principle of fertilisation, which, formalised in the meantime, required the egg to be fertilised by a spermatozoon, but also because they went against the common opinion of the 19th century, which could not conceive of sexual reproduction *sine concubitu*, i.e., without sexual intercourse. How could it be possible for an embryo to develop in the absence of a spermatozoon, considered to be the only possible cause of its development? The theory was ridiculed, foreshadowing, *mutatis mutandis*, what Charles Darwin would do a few years later with another theory that was even more heterodox, and of the immense scope that we know.

Fig. 13 Title page of the November 1845 issue of newspaper Bienen-Zeitung.

At the origin of the theory of parthenogenesis, which was later extended to all Hymenoptera, are the observations of the naturalist priest concerning the presence of "several young queens which, either because their wings were lame or because they were born during the cold season, had obviously not been able to take their nuptial flight and which, when dissected afterwards, were found not to be fertilised; they did, however, lay eggs from which only drones emerged. As these same eggs, developed in the ovary, would certainly have become worker eggs in the event of fertilisation, nothing could be more natural than to conclude that at the origin the eggs are all similar, or without distinction of sex, and that they become male or female according to whether they are deposited unfertilised or fertilised" (DZIERZON, 1882).

Jan Dzierzon knew about the existence and function of the spermatheca and that an egg can be fertilised or not by a spermatozoon. Had he read Karl von Siebold? The latter, probably on the basis of the work of Marcello Malpighi and Jean-Victor Audouin (see Chapter II), noted in 1837 the existence and role of the copulatory vesicle – the spermatheca – in female insects (queens and workers) and, in 1843, drew attention to this reservoir of sperm in the queens and workers of the genus *Apis* (SIEBOLD, 1857). Whether the act is voluntary or not, it is the release or not of a small quantity of sperm contained in the spermatheca as it passes an egg that determines fertilisation or the absence of fertilisation. We will return to this subject in the next chapter.

This discovery explains what neither Swammerdam, Réaumur nor even Huber had seen: A single egg is the origin of all three bee castes, and does not require any specialisation of the ovaries, contrary to what Adam Schirach had assumed.

This endless dispute about the reproduction of the Bees,

often carried on with great animosity, in which the opponents of

the different theories of generation relating to the Bees often

showed themselves to be mere dilettanti, miserably furnished

with natural-history information, was not fitted to attract the

interest of physiologists; indeed, it appeared as if the Apiarians

wished to fight the battle out amongst themselves without

foreign assistance, for the contest was never brought within

the province of an earnest investigation of nature. Moreover,

the naturalists could not very easily take part in the dispute,

as they were mostly deficient in the practical knowledge of

the oeconomy of Bees, without which every attempt to settle

the matter must have turned.

Karl von Siebold, "On a True Parthenogenesis in Moths and Bees" (SIEBOLD, 1857)

Jan Dzierzon battled alone for a while, but then a number of key figures came forward to support his theory, including Karl von Siebold, whom we have already met, Baron August von Berlepsch, a beekeeper from the Kingdom of Prussia, and the zoologist Rudolf Leuckart (see Figure 14).

Fig. 14 Top to bottom and right to left: Jan Dzierzon, Rudolf Leuckart, August von Berlepsch, Karl von Siebold (Die Gartenlaube journal, 1868, 597).

Before his meeting with Jan Dzierzon in 1851, Karl von Siebold was more opposed to parthenogenesis than in favour of it. Moreover, as can be seen from the above quotation, he did not hold beekeepers in high esteem, seeing them constantly quarrelling and remaining impervious to the knowledge accumulated by naturalists. However, his meeting with the Silesian pastor and the Prussian beekeeper, both of whom he held in high esteem, and his own subsequent research, made him the first scientific authority to rigorously prove the existence of parthenogenesis in psyches (a family of Lepidoptera), bees and silkworms (SIEBOLD, 1857). Jan Dzierzon described the scientist's contribution as follows: "The theory has passed the test of science under the microscope and the dissecting needles of the great physiologist, Professor Karl von Siebold. ... He undertook the tedious task of examining a number of worker eggs, as well as drone eggs, as to their fertilisation. The result was that the drone eggs showed no trace of fertilisation, either inside or out, while in the worker eggs one or more spermatozoa were clearly visible, some of which still showed vitality."

It was also Karl von Siebold who introduced the term parthenogenesis, which did not appear in the original expression of the hypothesis, and applied it to the phenomenon that Jan Dzierzon theorised.[19]

Jan Dzierzon also pointed out that he received his first swarm of Italian bees in 1853 and that the widespread distribution of this breed of bees in Europe, Scandinavia and the United States helped to speed up the recognition of his theory in a very simple and empirical way. Let us allow Baron August von Berlepsch to present the argument: "If the male egg does not need to be fertilised, pure Italian queens must invariably produce pure Italian males, and pure common queens must just as invariably produce pure common males, even if each race were fertilised by males of the other race. And this is the proven fact. Common queens pollinated by Italian drones are the clearest and most reliable proof. Among the drones produced by twenty common queens, in my apiary, impregnated by Italian drones and producing more or less mixed worker offspring, there is not one that bears the slightest resemblance to an Italian drone – all of them being perfectly of the common race" (BERLEPSCH, 1877). Of course, it would be different with queens from this first generation of queens.

Despite this elegant proof reported by Berlepsch, which many could see thanks to the growing presence of queens and therefore of males of the Italian breed, the controversy, which was initially expressed in the same Bavarian periodical *Bienen-Zeitung*, spread to other German and foreign publications.

Here is what Karl von Siebold regretfully noted in 1857, seven years after Jan Dzierzon's first publication:

> *Thus, in the various years of the Bienen-Zeitung up to the most recent period, we can find the following questions posed as not having a satisfactory answer, and the following points mentioned as doubtful by various beekeepers: Namely, whether the drones are really the male bees; whether the drones do not attend to the hatching of the eggs; whether the drones are not really abortions; whether there are not also male worker bees; whether the queen is not impregnated by caress or by mere agitation; whether copulation between the queen and a drone does not, after all, take place in the hive, and many other things besides.* (SIEBOLD, 1857)

---

19   This term was coined some time previously by the anatomist Richard Owen to designate another mode of reproduction. Until then, the established term was *Lucina sine concubitu* (*Lucina without sexual intercourse*) in reference to a satirical letter addressed by John Hill to the Royal Society of London which had refused his candidacy.

For his part, Baron August von Berlepsch, who like Karl von Siebold was initially an opponent of Dzierzon's theory, was converted and spared no effort to support the theory and its author. He called in Rudolph Leuckart, who dissected queens that had not mated and confirmed the absence of spermatozoa in their spermatheca (androtokia), but the presence of eggs in their ovaries. Thus, as Dr. Breyer said: "It is to Leuckart's credit that he provided the first direct demonstration and gave a naturalist's hypothesis an anatomical and physiological basis" (BREYER, 1868). In 1853 and 1854, Berlepsch published several letters, the "Apistical Letters," in the Bienen-Zeitung journal in particular. Later, in the first issue of American Bee Journal, in January 1861, he wrote a series of articles in which he presented and justified the theory (a summary of these articles was published in France by the journal L'Apiculteur in March 1862).

Always latecomers

Beekeepers and naturalists were more or less quickly convinced of the soundness of the theory, and some were never convinced at all. Here are a few examples.

In 1867, in France, a certain M. Landois[20] maintained that the sex of bees was solely the result of the quality of the food received by the larvae, and that this quality depended exclusively on the type of cell in which the larvae were found. Thus, according to this thesis, an egg initially laid in a male cell would have become a female if it had been transferred to a worker cell and vice-versa. It was the zoologist André Sanson and the pastor and beekeeper Frédéric Bastian who refuted this thesis at a meeting of the Académie des Sciences in Paris in 1868 (SANSON and BASTIAN, 1868).

In 1880 and 1881, i.e., 35 years after the 1845 publication, the columns of L'Apiculteur were still responding to the often vehement and discourteous criticisms levelled by the Italian clergyman and beekeeper Don Giotto Ulivi against the defenders of the theory of parthenogenesis, who were often contributors to this magazine as well as to American Bee Journal: "We defy to find a more chimerical, inconsistent and ridiculous system of reproduction than that of parthenogenesis" (ULIVI, 1880). In March 1881, in the same journal, he wrote: "Let the defenders of parthenogenesis make all the noise they want, let them imagine supports, let them publish their transcendental visions to the four winds, they will never succeed in destroying this natural and immutable order that the Creator has imprinted in every reproductive woman, that of being fertile only on condition of having been fertilised" (ULIVI, 1881). Following this logic,

---

20   It may be the German zoologist Hermann Landois (1835-1905).

he maintains that female workers are incapable of laying eggs (ULIVI, 1883).[21] He defended and relied on the thesis of the French professor of zoology Jean Pérez, who wrote: "The eggs of males, like the eggs of females, receive seminal baptism, and the theory of Dzierzon, created to explain a poorly observed fact, becomes useless if this [same] fact is disproved."

Jean Pérez, in a study unencumbered by any mockery, and in good faith unlike our Italian, contests Jan Dzierzon's theory. However, as curiously as paradoxically, he accepts it only in cases where the queen has not mated and in the case of the laying worker: "Drones can be born without a father; but if a father intervenes, he imbues them more or less strongly with the stamp of his race," he writes (PÉREZ, 1879).

Much later, in 1924, the same journal published an article entitled "The agony of Dzierzonnism." The author, a certain A. Bourgeois, wrote, not without a touch of arrogance: "I have demonstrated with irrefutable facts that the Immaculate Conception of males and the choice of the sex of the queen were a mere heresy on the part of scientists who had run out of arguments." He battled with all of them, including Charles and Camille Pierre Dadant who, he says, fiercely criticised him in the columns of American Bee Journal, L'Abeille du Canada and L'Apiculteur (BOURGEOIS, 1924). Admittedly, these are fine feats of arms (which, alas, he did not surrender) but they are also the sign of infinite stubbornness.

---

21   In this same text, Ulivi contests aerial fertilization ("invented by our Dreamer [François Huber] and supported by the imaginations of his too gullible and too affectionate disciples") and supports the possibility of a new fertilization of "old" queens (he claims to have seen it with his own eyes).

A history of the discoveries about the sexuality of the honey bee

Chapter VI
The choice of sex:
does the queen command nature?

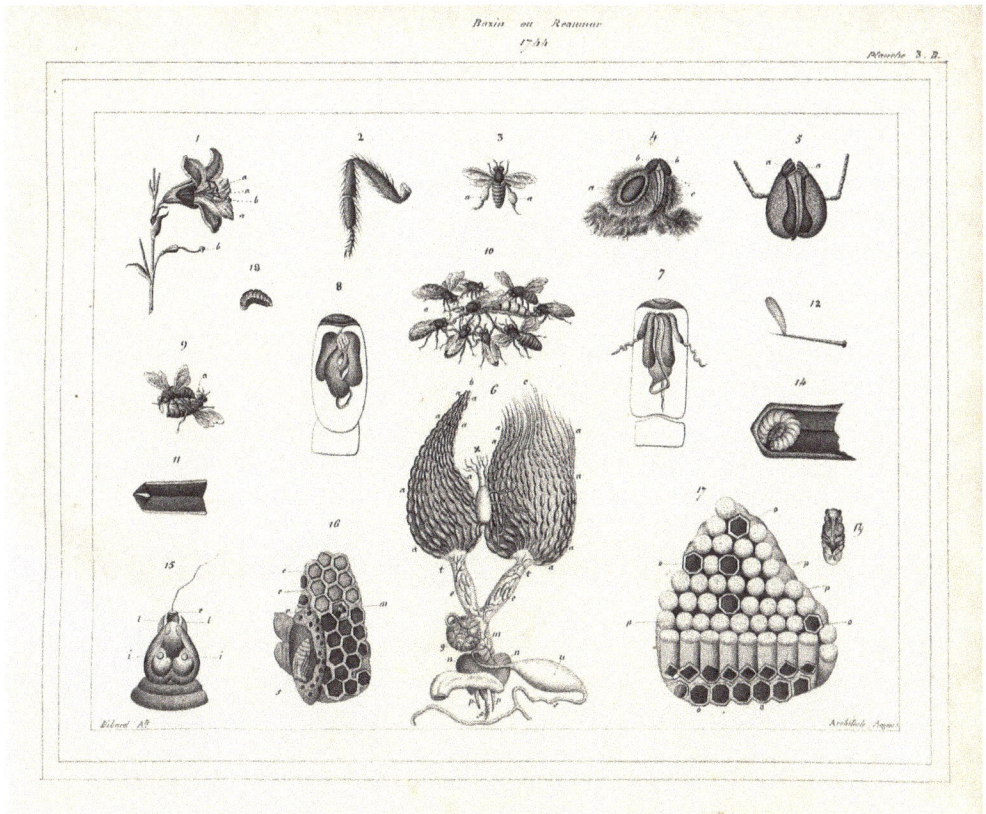

Fig. 15 Engravings from Réaumur (DEBEAUVOYS, 1855). In the center (G), the ovaries of a queen with the spermatheca (g).

If the queen's egg-laying can produce both male and female offspring, how is the sex determined? Does the queen have the power to choose the sex of her offspring? Réaumur thought so. He relates this anecdote:

> *"This bee seems to have some very singular knowledge, & knowledge that I have heard ladies envy it: They were shocked and complained that the mother bee seems to know what kind of bee should be born from the egg she is going to hatch, since she takes great care not to lay in a male cell, in a large cell, an egg from which only an ordinary bee should come, & that she never leaves in a small cell, in an ordinary cell, an egg that should produce a drone. The ladies I am talking about thought it wrong that nature should have taught simple bees so well, while leaving them unaware of the sex of the child they were to bring into the world".*
> (RÉAUMUR, 1740)

Jan Dzierzon is convinced that the choice of egg – fertilised or not – stems from the queen's own will. Anatomy seems to provide an answer in this sense. Thanks to an S-shaped valve located at the intersection of the copulatory vesicle – the spermatheca – and its duct, the queen could control whether or not a small quantity of sperm passed through and force the descent of this sperm toward the egg she was holding in front of the exit of the duct into the vagina. In addition, Karl von Siebold recognised that the spermatheca is surrounded by muscles which may be thought to be voluntary (DARESTE, 1858).

Another later theory, from the beginning of the 20th century, but no longer in use today, was that it was the bees that prevent fertilisation of the egg by the spermatozoa present on its surface, although it is not known how (MOREAUX, 1937). A variant of this theory was put forward by Charles Dadant, who believed that the distance between the royal legs relative to the size of the cell could play a role.

As for the so-called mechanistic thesis that the size of the alveolus has a mechanical effect by compressing the royal abdomen to a greater or lesser extent, this seems to have been refuted by the fact that the queen lays unfertilised eggs without fail in wide cells intended for rearing males and fertilised eggs in narrow cells intended for rearing workers or in a large cell such as a royal cell, even when these cells are barely stretched.

Finally, among the few other theories that have been expressed, we find that of the Italian Benedictine, Benussi-Bossi, who, in 1901, postulated that only the left ovarian oviduct would be in contact with the canal coming from the spermatheca. In this way, we return to Adam Schirach's thesis of the specialisation of the ovaries (see Chapter III), except that our monk does not say how the queen would "choose" to activate the oviduct according to the cell.

It is not certain that the question has been settled today.

Chapter VII
Lucina in the Sky with Drones[22]

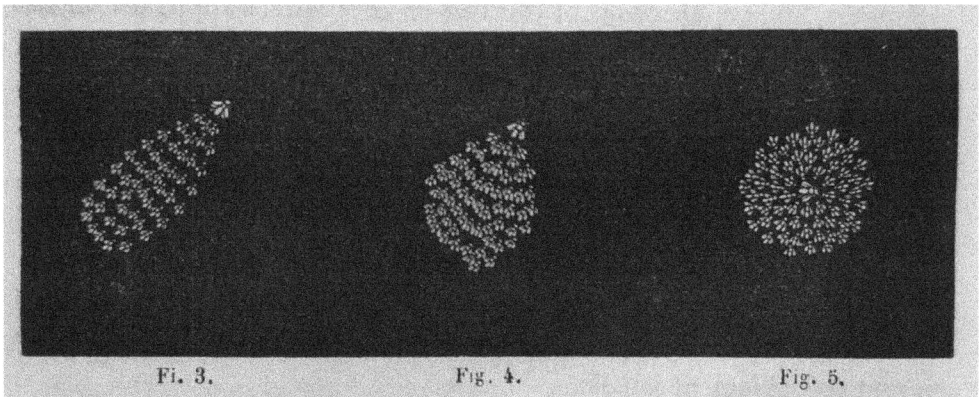

Fig. 16 Copulation of the mother bee, by Lorenzo Lorraine Langstroth
in L'Apiculteur of December 1861 (see Chapter IV).

Another great mystery of the beehive is the phenomenon known as polyandry,[23] in which the queen *Apis mellifera* mates with several males. In the 20th century, it became clear that this behaviour was, due to the genetic diversity it brings, the source of the honey bee major evolutionary traits.

How was this behaviour suspected, and then demonstrated, when queens could only be observed before and after one or more fertilisation flights and nothing was known about what happened in between? We have seen that, in the second half of the 18th century, copulation between the queen and a drone during a flight was established

---

22   Allusion to the Beatles' title of the song *Lucy in the Sky with Diamonds*. Lucina is the poetic epithet for Juno, Goddess of Light. She is invoked during childbirth and evokes marital fertility. See also note 19.

23   The term appears to have been in use since 1765 but is rarely found in books and articles on bees before the last few decades.

and partly described by François Huber and some of his precursors (see Chapter IV).

In his book "Vollständige Lehre von der Bienenzucht" *(Complete teaching of beekeeping)* published in 1775, two years after his death, Anton Janscha (JANSCHA, 1775) seems to mention the fertilisation of a queen by several males: in paragraph 8 devoted to the caste of drones, the author indicates that the queen "flies into the air (§. 6.) in order to be fertilised by the drones, after which she begins to lay eggs and will do so throughout her life." The use of the plural makes our aforementioned Slovenian scientist, Dr Salehar (see Chapter IV), say that Janscha is thus revealing the fact of multiple mating. However, as the author does not develop this important aspect any further, not even in paragraph 52 of the same opus, which deals with mating, it is doubtful that he really wanted to indicate that the queen is impregnated by several males during the same flight or during several successive flights.

François Huber notes in his second letter of 1791 that the queen is likely to undertake a second flight if the first has not resulted in a successful mating, i.e., when, for whatever reason and according to him, the eggs have not been fertilised (It should be remembered that François Huber envisages direct fertilisation of the eggs as soon as the sperm is emitted by a male – see Chapter III). He added: "It is very rare that a first mating is not sufficient" (HUBER, 1792). Surprisingly, it seems that no other author mentions multiple mating for a long time, which reinforces the idea that this fact had not been observed, or at least not by observers capable of reporting it.

Finally, a testimonial appeared in the columns of L'Apiculteur in October 1881. E. Pierrard, a beekeeper, described the return flights undertaken by queens and, more often than not, repeated over several days. The sign of fertilisation was generally present on the return flights (PIERRARD, 1881).

Some twenty years later, Amos Root, an American beekeeper and entrepreneur, mentions an accumulation of testimonies of multiple mating of the queen from 1904 onward (ROOT and ROOT, 1905), whereas, to his knowledge, nothing had been written on the subject since François Huber. He writes:

> Up till recently it was generally believed that the queen met the drone only once, I notwithstanding the fact that Francis Huber, in his book, "New Observations," published in 1814, made the statement that queens might or might not take more than one wedding-flight before beginning to lay. But this seems to have been overlooked until 1904, when considerable proof was adduced to show that the same queen before laying (not after) might not only take several wedding-flights, but come back on different occasions with the sure evidence of having met the drone.

In the August 1921 issue of L'Apiculteur, Marcel Gauthey, an abbot based in Saône-Et-Loire, a region of France, commented on Amos Root's book while observing differences in the livery of workers from a Caucasian queen in his own apiary. He suspected a possible double fertilisation by a male of the subspecies *caucasica* and a male of the subspecies *ligustica* (GAUTHEY, 1921). This article is supplemented by the account given by F. Julien de Mamers in the same journal in 1922 (MAMERS, 1922) and 1924 (MAMERS, 1924). In 1924, he concludes his observations as follows: "Admitting that a queen is only fertilised once in her life (which I do not absolutely deny), it is neither in a single day, nor by a single male nor by a single mating, since the least demanding go out two days in a row and come back each time with a male bulb. This is how five out of the seven experienced subjects behaved, not counting the other two who went out and reported signs of mating three and four days in a row. I would like to point out that not one of these seven queens was completely fertilised, either on the first day or by a single mating."

The Frenchman Adrien Perret-Maisonneuve also mentioned multiple fertilisation in 1923 (PERRET-MAISONNEUVE, 1923).

And finally, in 1935, Ernest Root, son of Amos Root, reported two eyewitness accounts published in the American magazine Gleanings of Bee Culture in July and November 1927 (ROOT and ROOT, 1935). These two eyewitness accounts report a succession of two fertilisation flights with, on each return, what François Huber called signs of fertilisation, visible at the tip of the royal abdomen (NELSON, 1927; CORYELL, 1927).

Since François Huber, the sign of fertilisation or mating was thought to be a part of the bulb of the endophallus embedded in a mass of sperm, but Jerzy Woyke and Friedrich Ruttner (WOYKE and RUTTNER, 1958) demonstrated that these were chitinous plates, mucus and glandular epithelium released from the endophallus without it being, strictly speaking, sectioned or torn (WOYKE, 2010). In fact, it is the glandular epithelium, which can take the form of a white filament, that can sometimes be seen with the naked eye at the end of the queen's open abdomen on her return. The mucus is always expelled from the male organ after the sperm, which is lodged in the oviduct before a smaller fraction reaches the spermatheca. It has been shown that although a drone produces a quantity of sperm greater than the capacity of the spermatheca, most of the ejaculate, estimated at 97%, does not reach the spermatheca (KOENIGER and KOENIGER, 2000) and is therefore lost.

The function of the mating sign has been the subject of several hypotheses. One hypothesis is that it serves to prevent sperm flow after mating, as mucus coagulates when exposed to air (BISHOP, 1920; WOYCIECHOWSHI, 1994). Another hypothesis is that, contrary to what happens in many species in which mechanisms or behaviour prevent or hinder subsequent copulation with another male, this sign would be attractive to other males, signalling more explicitly a nubile female (KOENIGER, 1990).

In June 1940 in the United States, William C. Roberts undertook a study on the fertilisation of 40 virgin queens (ROBERTS, 1944). In a fertilisation station containing males of the local subspecies *mellifera* and *ligustica*, the author observed the number and duration of fertilisation flights, as well as the colour of the abdominal segments of the offspring of each queen. However, this last categorisation does not seem to be related to the number of flights (one or two, rarely three). This observation of a mixed progeny for a single fertilisation flight would, however, have made it possible to demonstrate multiple mating during a single flight.

Until the early 1950s, the possibility of a succession of matings during a single flight was not raised, but the decade ahead will bring answers: Jerzy Woyke, a Polish professor of agricultural science specialising in beekeeping, credited V.V. Triasko (1951), Stephen Taber (1954) and himself (1955) with providing evidence of multiple mating during the same fertilisation flight.

How did these researchers go about it?

Triasko, a Russian biologist, began dissecting queens returning from their first fertilisation flight and compared the volume of sperm present in the oviducts with the volume contained in the seminal vesicles of a male. The researcher then estimated that each queen had mated on average with 4-5 drones and also assumed that each drone removed the mating sign of its predecessor in order to mate (TABER, 1954). Taber calculated an average number of matings of six, based on the frequency of appearance of a specific mutation in the progeny of queens fertilised in a fertilisation station previously saturated with two types of drones, some of which carry the mutation (TABER, 1954; PEER, 1956).

Woyke proved this by studying a large number of mating signs (WOYKE, 1955). This proof is materialised by the fine orange membranes found on both sides of the mating sign. These membranes come from the horns of a male's endophallus and detach when these horns enclose the mating sign of a predecessor when it is present. Thus, the male whose orange membranes are found is the male who was unable to remove the mating sign of the previous male (WOYKE, 2011). Professor Karl von Siebold noted this in 1854, but the observation was not confirmed in subsequent publications. Although queens may return from a fertilisation flight with no visible sign of mating, even though they have been fertilised (WOYKE, 1956), a large proportion return with a visible sign of mating at the tip of the abdomen. This sign generally shows the orange membranes from the male – more rarely from two males – that failed to remove the mating sign from a predecessor.

In most cases, the presence of a mating sign does not prevent a new coitus, as the organ of another male may be introduced between it and the bottom of the sting chamber, along the last sternite. The mating sign of a predecessor is generally expelled during the second and final phase of complete eversion of the endophallus, which is the irreversible turning over of the organ, previously entirely contained in the abdomen, toward the outside.

We now know that the number of mating events during a single flight is highly variable. They follow each other in quick succession, each lasting no more than 5 seconds, which leaves little room for the hypothesis that the queen is exercising any kind of choice. However, the question of whether and how the queen exercises any kind of choice remains open, whether during the flight or afterwards, when other mechanisms could come into play (BAER, 2005).

It seems that a flight is interrupted when a new mating does not result in the removal of a sign of mating. The queen then returns to the nest and, on the same day or the following day, undertakes a new fertilisation flight after she, alone or with the help of the workers in her swarm, has managed to remove the mating sign. A subsequent flight seems to be correlated more with an insufficient number of matings than with the quantity of sperm collected (SCHLÜNS et al., 2005)

More recently, new estimates of the number of matings based on molecular genotyping of a queen's offspring range from 1 to 59 and average around 14, regardless of the number of flights (TARPY et al., 2015).

To conclude this chapter, the existence of polyandry was demonstrated without anyone ever having directly observed a series of mating. This did not, however, reduce the number of questions raised, as new and fascinating questions arose that many studies have explored in recent decades. To begin with, why has this strategy evolved? What are the effects of the genetic diversity it generates on the maintenance and survival of a colony in changing environments – in other words, what is its adaptive value? If polyandry leads to a reduction in kinship within the colony, is it a source of conflict, discrimination or nepotism? What effects does it have on cooperative behaviour? Why are so many males produced when very few will mate with a queen? There are many more questions to be answered.

A history of the discoveries about the sexuality of the honey bee

Chapter VIII
Happy ending?

Fig. 17 Mating of a queen and a male *Apis mellifera*.
At the end of the mating, the drone falls and the queen regains speed and altitude.
Drawing by Fany Eggers, courtesy of the author.

Sex and death

As we know, the male does not survive mating, but it is not easy to find the moment or the author to whom knowledge of this fact should be attributed. Réaumur and Huber both mentioned the instantaneous death of the drone, once the eversion of its endophallus had been provoked by pressure on its abdomen, but also the natural death of some drones after spontaneous eversion, not provoked by pressure. However, neither of them had looked at or laid hands on a drone as soon as natural mating had taken place.

A touch of anthropomorphism and a cheap moral lesson

The former French lawyer, revolutionary commissioner and beekeeper Charles Romain Féburier, commenting on Réaumur's unsuccessful attempts to induce mating between a queen and a drone (see Chapter II), refers to the death of the drone and the latter's possible premonition of it:

> It is their [the males'] cold temperament that forces the mother bee to make all the advances; it seems to animate them only with difficulty. It seems that these males foresee their fate after copulation, and that the fear of death, which for them is the necessary consequence of copulation, since in separating from the mother-bee they leave in her vulva the constituent parts of their being, makes them very reluctant to fulfil the end of their destination; but this is pushing the foresight of these insects too far, and Réaumur's experiment does not seem to me to be conclusive enough. (FÉBURIER, 1810)

In 1790, in Volume 2 of his "Traité complet sur les abeilles" (*Complete treatise on bees*), Abbé Della Rocca, vicar of the Greek island of Syros, also recalled Réaumur's experiments, but took the opportunity to deliver a moral lesson: "If we could lend reasoning to these males, we would say that they are not very wrong to resist this action [the queen's solicitations], and not to succumb to it first; for, according to M. de Réaumur's remark, they die soon after mating, so that this pleasure of a moment costs them dearly. A good lesson for libertines" (DELLA ROCCA [Abbé], 1790).

Be that as it may, this end of the drone seems so unfortunate that our curiosity, perhaps a little compassionate, leads us to ask human questions, all too human: How does the male live through this experience? We obviously don't know (nor do we know anything about the queen's subjective experience). Can the male avoid the perils of mating? Does he have a choice? No, probably not. We can assume that he is motivated and constrained by a powerful and irrepressible biological imperative and, to put it trivially (but wrongly if we exclude any idea of finalism), this is the whole purpose of his existence and, indeed, its end.

A fatal reversal

We often read that, as soon as mating begins, the male is paralysed from the head to the thorax. If this were the case, wouldn't he lose his ability to fly and direct his flight? This seems to be what happens, but shortly before the pair separate. Perhaps this cessation of wing beating marks the moment of the male's death? Whatever the case, the muscles of the abdomen remain free and their contractions exert great pressure on the haemolymph, allowing the endophallus to be everted outwards in several phases. Jerzy Woyke fixes the moment of death at the last stage of endophallus eversion, when abdominal pressure is at its maximum and the male's semen is emitted with force (WOYKE, 2008). However, some authors claim that it breathes its last later after an agony lasting from a few minutes to several hours.

At the end of the eversion, the size of the male's abdomen is said to be greatly reduced (LANGSTROTH, 1861 ; KOENIGER et al., 2014), down to a quarter of its initial size (WOYKE, 2016). Karl von Siebold attributes to Rudolph Leuckart the idea that flight allows the respiratory system to be filled to capacity, which he thought was a necessary prerequisite for the complete eversion of the male organ (SIEBOLD, 1857). This idea can also be found in the work of Michel Girdwoyn (1876).

What exactly does the male die of? exhaustion? A large transfer of haemolymph to the endophallus? Damage to his organs as a result of ejaculation? The characteristic sound witnessed by some observers reflects the force required to provoke eversion. This sound has been compared to that produced by a roasted grain of wheat. It would appear that there are no studies on the precise cause of death. It should be pointed out here that eversion of the endophallus is irreversible, making the possible survival of the male, like Priape the Greek god of fertility, untenable to say the least. Finally, a risky speculation: Would the queen use her sting to dismiss her lover? It does not seem that this last speculation was ever put forward.

A final cheap moral lesson, but a funny one

To close this chapter and this detour in an attempt to piece together the story of the discoveries about the honey bee sexuality, leaving the drone to its fate, let us recall an anecdote inspired by its end *in coitus*, which appears in the Gazette Médicale de Paris of 1854: "The royal coupling has only been seen once since the beginning of the centuries, and it was a Trappist priest from the Melleray Abbey in Brittany, a region of France, whom Providence chose to witness it; a wise choice if ever there was one, for there was no better way than by such an example to prove to the reverend father that chastity is better than love" (JACQUOT, 1854).

Acknowledgments

John Murphy for his assistance in translating several passages in 18th and 19th-centuries German from the works of Anton Janscha and Jan Dzierzon.

Clotilde Randriamampita for her attentive proofreading.

Fany Eggers de Villepin for her drawings.

Bibliography

BAER, B. 2005. Sexual selection in *Apis* bees. *Apidologie*. 36:187–200. doi:10.1051/apido:2005013.

BERLEPSCH, A. (Baron de). 1877. The Dzierzon theory; being a full elucidation of scientific bee-keeping by the Baron of Berlepsch [02509]. American Bee Journal.

BISHOP, Geo.H. 1920. Fertilization in the honey-bee. I. The male sexual organs: Their histological structure and physiological functioning. *J. Exp. Zool.* 31:224–265. doi:10.1002/jez.1400310203.

BOURGEOIS, A. 1924. L'agonie du Dzierzonnisme. *L'Apiculteur*. LXVIII:123–125.

BREYER, A. 1868. À propos de la thèse sur la parthénogénèse, soutenue par M. Plateau prof. à l'Athénée de Bruges, devant la Faculté de Gand. *Ann. Société Entomol. Belg.* XII:XXII–XL.

BUTLER, C. 1609. The Feminine Monarchy or the History of Bees. London.

CLUTIUM, T. 1597. Van De Byer. Amsterdam.

COBB, M. 2002. Jan Swammerdam on social insects: a view from the seventeenth century. *Insectes Sociaux*. 49:92–97. doi:10.1007/s00040-002-8285-z.

CORYELL, R.M. 1927. QUEEN MATES TWICE. *Glean. Bee Cult.* 55:443.

DARESTE, C. 1858. La Parthénogénèse. Revue germanique. 5 pp.

DEBEAUVOYS, C.P. 1855. Dictionnaire d'apiculture. Manuscrit. 2500 pp.

DEBRAW, J. 1778. Debraw's Difcoveries on the Sexes of Bees. *Gentlem. Mag. Hist. Chron.* 48:215–218.

DELLA ROCCA (Abbé). 1790. Traité complet sur les abeilles, avec une méthode nouvelle de les gouverner telle qu'elle se pratique à SYRA ... 2. IParis, mp. de Monsieur.

DZIERZON, J. 1845. Gutachten über die von Hrn. Direktor Stöhr im ersten und zweiten Kapitel des General-Gutachtens aufgestellten Fragen. *Bienenztg. Hrsg. Von Dr C Barth Schmid Eichstädt Jahrg. I.* 109–113, 119–122.

DZIERZON, J. 1882. Dzierzon's rational bee-keeping, or, The theory and practice of Dr.

Dzierzon. Houlston & sons , London.

FÉBURIER, C.R. 1810. Traité complet théorique et pratique sur les abeilles. Paris, Imp. de Mme Huzard. IV–460, 1 pl. dépl. h.-t. pp.

GAUTHEY, M. 1921. Une reine peut-elle être fécondée deux fois ? *L'apiculteur*. 185-189.

GIRDWOYN, M. 1876. Anatomie et physiologie de l'abeille. 1. Paris, Rouge, Dunon et Fresné. 32, 12 pl. pp.

GUZMÁN-ÁLAVAREZ, J.R. 2006. Tratado breve de la cultivación y cura de las colmenas … Edición crítica de José Ramón Guzmán Álvarez [50110].

HUBER, F. 1792. Nouvelles observations sur les abeilles à adressées M. Charles BONNET. À Genève : Chez Barde, Manget & Compagnie, Imprimeurs-Libraires.

HUBER, F. 1814. Nouvelles observations sur les abeilles. 2. 2nd ed. Paris, chez J.-J. Paschoud, lib. 479, 12 pl. pp.

JACQUOT, F. 1854. Exposition permanente des produits de l'Algérie et Concours Général Agricole de 1854 [50101]. *Gaz. Médicale Paris*. XXIV ANNÉE. TROISIÈME SÉRIE, TOME IX.

JANSCHA, A. 1775. Vollständige Lehre von der Bienenzucht (Enseignement complet de l'apiculture). Wien.

KOENIGER, G. 1990. The role of the mating sign in honey bees, Apis mellifera L.: does it hinder or promote multiple mating? *Anim. Behav.* 39:444-449. doi:10.1016/S0003-3472(05)80407-5.

KOENIGER, G., N. KOENIGER, J.D. Ellis, and L. Connor. 2014. Mating biology of honey bees (*Apis mellifera*). Wicwas Press.

KOENIGER, N., and G. KOENIGER. 2000. Reproductive isolation among species of the genus Apis. *Apidologie*. 31:313-339. doi:10.1051/apido:2000125.

LANGSTROTH, L.L. 1861. Copulation de l'abeille-mère. *L'apiculteur*. 6 (1861-1862):79-83.

LERICHE, J.-B. 1884. Rectifications sur rectifications relatives aux ouvrières pondeuses. *Bull. Société Apic. Somme*.

LEUCKART, R. 1861. The sexuality of bees by Professor LEUCKART. *Am. Bee J.* 1.

LOMBARD, C.-P. 1805. État de nos connaissances sur les abeilles au commencement du XIXe siècle. Paris, lib. de Mme Huzard. 72 pp.

MALPIGHI, M. 1686. La Structure du ver a soye et de la formation du poulet dans l'œuf contenant deux dissertations de Malpighi. À Paris chez Maurice VILLEYY.

MAMERS, F.J. (de). 1922. Une reine peut-elle être fécondée deux fois avant de commencer sa ponte ? *L'Apiculteur*. LXVI:44-45.

MAMERS, F.J. (de). 1924. La fécondation des reines. *L'Apiculteur*. LXVIII:138-142.

MÉNDEZ DE TORRES, L. 1586. Tractado breue de la cultiuaci☐ y cura de las colmenas. Alcala, en la caía de Juan Iñienez de Lequerica.

MOREAUX, R. 1937. Sur le déterminisme du sexe chez l'abeille. *Gaz. Apic*. 38.

NELSON, F.C. 1927. DOUBLE QUEEN-MATING. *Glean. Bee Cult*. 55:720-721.

PEER, D.F. 1956. Multiple Mating of Queen Honey Bees. *J. Econ. Entomol*. 49:741-743. doi:10.1093/jee/49.6.741.

PÉREZ, J. 1879. Mémoire sur la ponte de la reine et la théorie de DZIERZON. 22 pp.

PERRET-MAISONNEUVE, A. 1923. L'Apiculture intensive et l'élevage des reines. Paris, Maurice Mendel, édit. 450, 70 fig. pp.

PIERRARD, E. 1881. Observations sur la fécondation. *L'Apiculteur*. 295-300.

RÉAUMUR, R.A.F. (de). 1740. Mémoire pour servir à l'histoire des insectes. 5. À Paris, de l'Imprimerie Royale.

REMNANT, R. 1637. A discourse or historie of bees Shewing their nature and usage, and the great profit of them. LONDON, Printed by Robert Young for Thomas Slater, dwelling in duck lane at the white Swan.

ROBADEY, V. 2021. Les circulations et les transferts agronomiques entre la Société économique de Berne et les sociétés d'agriculture françaises (1757-1773): l'exemple des abeilles et de l'Encyclopédie économique.

ROBERTS, W.C. 1944. Multiple mating of queen bees proved by progeny and flight tests. *Glean. Bee Cult*. 72:255-259, 303.

ROOT, A.I., and E.R. ROOT. 1905. The ABC of Bee Culture. Medina, Ohio, USA, A.J. Root Co. 490, ill. pp.

ROOT, A.I., and E.R. ROOT. 1935. The ABC and XYZ of Bee Culture. The A.I. ROOT Company, Medina, Ohio.

SALEHAR, D.A., and F. SIVIC. 2021. Records (manuscripts and publications) on the mating of the queen with drones in the air in the years 1763 - 1776 from Slovenia.

SANSON, A., and F. BASTIAN. 1868. Expériences sur la transposition des œufs d'abeilles au point de vue des conditions déterminantes des sexes [01123]. Académie des sciences. 6 pp.

SCHIRACH, A.G. 1761. Die mit Natur und Kunst verknüpfte neuerfundene Oberlausitzsche Bienen-Vermehrung, oder Junge Bienen-Schwärme beym Anfange des May-Monats in Wohnstuben zu machen : Nebst andern nützlichen und erbaulichen Anmerkungen von Bienen. Richter, David, Budißin.

SCHIRACH, A.G. 1770. Ausführliche Erläuterung der unschätzbahren Kunst, junge Bienenschwärme, oder Ableger zu erzielen : nebst einer natürlichen Geschichte der Bienenkönigin ...

SCHIRACH, A.G. 1771. Histoire naturelle de la reine des abeilles, avec l'art de former des essaims. On y a ajouté la correspondance de l'auteur avec quelques savants, & 3 mémoires de l'illustre M. Bonnet de Genève sur ses découvertes. La Haye, Pays-Bas, Frédéric Staatman. LV-269 pp.

SCHLÜNS, H., R.F.A. MORITZ, P. NEUMANN, P. KRYGER, and G. KOENIGER. 2005. Multiple nuptial flights, sperm transfer and the evolution of extreme polyandry in honeybee queens. *Anim. Behav.* 70:125-131. doi:10.1016/j.anbehav.2004.11.005.

SCOPOLI, G.A. 1763. Entomologica Carniolica exhibens insecta Carnioliae indigena et distributa in ordines, genera, species, varietates methodo Linnaeana. Vindobonae, Trattner, 1763.

SIEBOLD, K.T.E. 1857. On a True Parthenogenesis in Moths and Bees. John van Voorst. Paternoster Row., London.

SIMON, J.B. 1742. Le gouvernement admirable ou la république des abeilles. 2nd ed. Paris, Chez Thiboust. LXIV-390, 5 pl. h.-t. pp.

SWAMMERDAM, J. 1682. Histoire générale des insectes. Utrecht, Pays-Bas, chez Guillaume de Walcheren, lib . 6-215, 13 pl. in-t.1 tableau pp.

SWAMMERDAM, J. 1737. Bybel der natuure - Biblia naturae. 1. Te Leyden, by Isaak Severinus, Boudewyn van der Aa, Pieter van der Aa.

SWAMMERDAM, J. 1758. Histoire naturelle des insectes. Dijon, Paris, 673 p. & 36 planches hors texte.

TABER, S., III. 1954. The Frequency of Multiple Mating of Queen Honey Bees. *J. Econ. Entomol.* 47:995-998. doi:10.1093/jee/47.6.995.

TARPY, D., D.A. DELANEY, and T.D. SEELEY. 2015. Mating frequencies of honey bee queens (Apis mellifera L.) in a population of feral colonies in the Northeastern United States. *PloS One*. 10:e0118734. doi:10.1371/journal.pone.0118734.

TRIASKO, V.V. 1951. Priznaki osemiennosti pchelinych matok. [Signs of queens' mating] Article non retrouvé. *Pchelovodstvo*. 28:25–31.

ULIVI, G. 1880. Réponse aux critiques de Monsieur VIENNEY, collab. du journal l'Apiculteur de Paris. Turin, Italie, Etablissement artistique-Littéraire. 22 pp.

ULIVI, G. 1881. Réplique Ulivi sur la fécondation. *L'Apiculteur*. 25:82–88.

ULIVI, G. 1883. Les vieux croyants, ou les abeilles tutrices confondues avec les abeilles ouvrières [01193]. Établissement Artistique- Littéraire. 14 pp.

VOLTAIRE. 1789. Dictionnaire philosophique. 1. Nouvelle édition. Marc-Michel Rey, Amsterdam.

WOYCIECHOWSHI, M. 1994. The function of the mating sign in honey bees.

WOYKE, J. 1955. Multiple Mating of the Honeybee Queen (Apis mellifica L.) in One Nuptial Flight. *Bull Acad Pol. Sci Cl II*. 3:175–180.

WOYKE, J. 1956. Anatomo-Physiological Changes in Queen-Bees Returning from Mating Flights, and the Process of Multiple Mating. *Bull. L'ACADBMIE Pol. Sci. Cl II*. IV.

WOYKE, J. 2008. Why the eversion of the endophallus of honey bee drone stops at the partly everted stage and significance of this. *Apidologie*. 39:627–636. doi:10.1051/apido:2008046.

WOYKE, J. 2010. Three substances ejected by Apis mellifera drones from everted endophallus and during natural matings with queen bees. *Apidologie*. 41:613–621. doi:10.1051/apido/20010007.

WOYKE, J. 2011. The mating sign of queen bees originates from two drones and the process of multiple mating in honey bees. *J. Apic. Res.* 50:272–283. doi:10.3896/IBRA.1.50.4.04.

WOYKE, J. 2016. Not the Honey Bee (*Apis mellifera* L.) Queen, but the Drone Determines the Termination of the Nuptial Flight and the Onset of Oviposition - The Polemics, Abnegations, Corrections and Supplements. *J. Apic. Sci.* 60:25–40. doi:10.1515/jas-2016-0032.

WOYKE, J., and F. RUTTNER. 1958. An Anatomical Study of the Mating Process in the Honeybee. *Bee World*. 39:3–18. doi:10.1080/0005772X.1958.11095028.

A history of the discoveries about the sexuality of the honey bee

www.ingramcontent.com/pod-product-compliance
Lightning Source LLC
Chambersburg PA
CBHW061457270326
41931CB00021BA/3487